Ethical Questions in Brain and Behavior

Problems and Opportunities

Ethical Questions in Brain and Behavior

Problems and Opportunities

Edited by
Donald W. Pfaff

Springer-Verlag
New York Berlin Heidelberg Tokyo

Donald W. Pfaff
Laboratory of Neurobiology and Behavior
The Rockefeller University
1230 York Avenue
New York, New York 10021
U.S.A.

Library of Congress Cataloging in Publication Data
Main entry under title:
Ethical questions in brain and behavior.
  Bibliography: p.
  Includes index.
  1. Psychiatric ethics.   2. Critical care medicine—
Moral and ethical aspects.   I. Pfaff, Donald W.,
1939–   . [DNLM: 1. Psychophysiology.   2. Mental
disorders.   3. Nervous system diseases.   4. Ethics,
Medical. WL 103 E84]
RC455.2.E8E834   1983   174'.2   83-14586

With 3 Figures.

Typeset by Ampersand, Inc., Rutland, Vermont.
Printed and bound by R.R. Donnelley & Sons, Harrisonburg, Virginia.
Printed in the United States of America.

9 8 7 6 5 4 3 2 1

ISBN 0-387-90870-6 Springer-Verlag New York Berlin Heidelberg Tokyo
ISBN 3-540-90870-6 Springer-Verlag Berlin Heidelberg New York Tokyo

Contents

Contributors

COLIN BEER
Institute of Animal Behavior, Rutgers University, Newark, New Jersey 07102 U.S.A.

H. RICHARD BERESFORD
Department of Neurology, North Shore University Hospital, Manhassett, New York 11030 and Cornell University Medical College, New York, New York 10021 U.S.A.

JERRAM L. BROWN
Department of Biological Sciences, State University of New York, Albany, New York 12222 U.S.A.

ARTHUR L. CAPLAN
The Hastings Center, Hastings-on-the-Hudson, New York 10706 U.S.A.

JEAN ENDICOTT
Department of Psychiatry, College of Physicians and Surgeons, Columbia University, New York, New York 10032 U.S.A.

DAVID E. LEVY
Department of Neurology, Cornell University Medical College, New York, New York 10021 U.S.A.

RUTH MACKLIN   Department of Community Health,
                       Albert Einstein College of Medicine, Bronx,
                       New York 10461 U.S.A.

DONALD W. PFAFF   Laboratory of Neurobiology and Behavior,
                       The Rockefeller University, New York,
                       New York 10021 U.S.A.

FRED PLUM   Chairman, Department of Neurology,
                       Cornell University Medical College,
                       New York, New York 10021 U.S.A.

Chapter 1

Introductory Remarks

FRED PLUM, M.D.,* AND DONALD PFAFF, PH.D.**

Clinical Perspective

Webster's dictionary defines ethics as "a group of moral principles or set of values . . . governing the behavior of an individual or a profession." A moment's reflection reminds one that the definition carries inherent anthropological limits: Most ethical principles represent tribal rules, not guidelines universally deduced by the species as being to their advantage. By no means can we prove that Fiji Islanders necessarily agree with Kant's categorical imperatives. Furthermore, even within our Western tribes, subcultures differ vigorously in some aspects of their ethics. Witness the divisions that separate the principles of the Moral Majority from what many academics believe. What is it then that stimulates the discussions summarized in this book? Certainly it is not any confidence in a zealously held set of ethics that any of the authors wish to impose on others. Rather, my own reasons (F.P.) lie in the fact that one can hardly be a doctor without subjecting oneself to a good deal of questioning about one's own and one's associates' behavioral principles. As a result, as Paul Ramsey well states: "The Nuremberg Code, the Declaration of Helsinki, various 'guidelines' of the American Medical Association and other codes governing medical practice constitute a sort of catechism in the ethics

*Department of Neurology, Cornell Medical College, New York, New York 10021.
**Laboratory of Neurobiology and Behavior, The Rockefellar University, New York, New York 10021.

of the medical profession. These codes exhibit a professional ethics which ministers and theologians and members of other professions can only profoundly respect and admire." Nevertheless, as Ramsey goes on to point out, a catechism has never alone sufficed to guide behavior. Unless these established ethical principles are constantly pondered and enlivened in their application, they become dead letters. Furthermore, there is the need to deepen our principles, to sensitize and open to further humane revision in the face of all the ordinary and newly emerging situations which a doctor confronts. In this task, no sources of moral insight, no articulate understandings of the humanity of man or for answering questions of the medical management of life and death can rightfully be neglected.

More specific reasons, however, than merely the drives of constant introspection and a desire for a moral "trading up" have stimulated the present course. Advances in medicine during the past 30 years have resulted in profound changes of human life. We live substantially longer than our fathers, and medicine now possesses the capacity to alter drastically the function of almost every organ in the body, including the brain. Furthermore, present scientific advances in neurobiology are such that we expect that the capacity to manipulate the functions of mind and brain will be even greater in the future than it is today. Under the circumstances, it is time to take stock and begin to plan for the future. Below we allude to the problems posed when psychiatric disease impairs so greatly the "organ of consent" that the person cannot give permission for his own necessary treatment. Severe brain disease in the unborn and postnatal child as well as in the adult imposes equally great ethical problems about when and what kind of treatment is best applied. While it is not the privilege of medicine alone to suggest the ethical solutions to these issues, only medical science can provide the factual evidence which society requires in order to construct its ethical guidelines.

If one accepts that all we know, we know through our brain and that much of our social behavior reflects the efforts of this biological organ to govern our behavior for maximal reward and survival in a potentially hostile environment, then morality and ethics perhaps can be viewed as appropriate objects of neurobiological and neurological examination. At the very least, the fundamental neurobiologist can examine individual morality and group ethics abstractly, as he does other forms of behavior, determining what is the prevailing norm and asking how the brain regulates organisms to produce that norm. At the same time, the experience of the clinical neurologist leads him to ask what impact modern medicine with its remarkable capacity to extend the natural life and function of almost every organ *but* the brain may have upon ancient ethical principles. Most of the latter evolved during times when disease in other organs almost always produced death *before* the brain had a chance to fail.

The success of medicine in extending the life or function of almost every nonneurological organ of the body undoubtedly is a major factor that has led to the extensive questioning of medical ethics that one today encounters in so many quarters of society. For most of the history of our race until those alive at this moment, the failure of one or more systemic organs limited the duration of life's span long before the full capacities of the thinking brain had run its course. The surpassing ethical principles of free will, individual autonomy, self-determination, and the sanctity of life grew from minds that, in a sense, could not only anticipate their final days but also usually lived them out possessed of a sufficient lucidity to watch the withering of other parts of their bodies that predicted their own death. Probably nothing better illustrates modern man's departure from the days of these archaic principles than the present widespread acceptance of the ethic that death of the brain means death of the person. This acceptance overturns a deeply rooted belief which during all former ages held that the human spirit departed only with the cessation of heart action or breathing. Nowadays, technology allows almost every organ except the brain to have its natural functions extended or replaced, in many instances for a long period of time. This startling change leaves mankind increasingly uncomfortable about how to apply classic ethical principles to bodies which have lost much or all of the capacity for making decisions about self-determination, but nevertheless will survive for long periods so long as attentive care is provided. Stating the question directly: How does one obtain informed consent from an organ so disordered or damaged that it lacks and never will regain its normal capacity to judge itself?

Our chapters draw upon only a small fraction of thoughtful Western writings which have tried to deal with ethical principles that impinge upon medicine. Certain issues or questions arise repeatedly, however, and these are set out as alternatives in the following paragraphs.

1. Quality of life versus sanctity of life. The ethical choice here centers on whether or not the physician has the right to limit treatment when the patient's condition is hopeless and pain or other overwhelming limitations have destroyed the individual's every desire to survive. Sharp divisions of opinion exist, with some going so far as to liken the alternatives as epitomizing differences between stoic and Judeo–Christian philosophy; we find that too strongly stated a contrast and perhaps an implicitly pejorative way of expressing the issue. There is no question, however, but that some ethical thinkers as well as certain legal minds envision any reduction of care by the physician as creating the risk of a first dangerous step toward administratively sanctioned deprivations of individual rights. Most doctors disagree, although they recognize their own complicated feelings and biases that influence the dilemma of when and whether to deliver merciful comfort rather than to extend the agony of pain-wracked terminal survival.

2. *Rights of the individual versus rights of the society.* This ethical choice arises repeatedly in both financial and social cost–benefit decisions. For example, does one employ highly expensive technology such as renal dialysis under all medically appropriate circumstances or does one limit such treatments according to considerations such as the age and general health of the patient. The ethical alternatives also directly arise in questions concerning whether and what kind of treatment should be applied to greatly extending the lives of the severely brain damaged. A less obvious part of this general heading has to do with issues of how to balance the rights or needs of the already sick versus those who do not yet require medical care. This is a major issue, for example, when it comes to judging measures which might control reproduction in persons carrying major genetic diseases. It also arises when one makes decisions to transfer severely but hopelessly ill patients from acute care hospitals to custodial care institutions so that severely ill patients, as yet unnamed, can take advantage of the acute units.

3. *The sanctity of life versus personal autonomy.* Does the individual have the right to die when he so chooses, and should living wills be binding?

4. *Personal autonomy versus medical need.* This issue arises in relation to the prohibition of certain sects against receiving blood, the refusal of certain patients with severe psychiatric disease to accept treatment that would relieve their symptoms or assure their safety, and the capacity to treat in emergency circumstances those who have not given informed consent. The same issue emerges in questions as to whether and when medical paternalism should influence or transcend the process of "informed consent."

5. *The domain of the court versus the domain of medicine or even the domain of the individual.* Society increasingly questions the privilege of medicine in making ethical decisions that extend beyond mere technical knowledge. Conversely, courts recently have made decisions which imply greater possession of specialized technical knowledge and its medical application than some would have thought appropriate to their education or mandate. What is to be the impact of these twin changes on our ethic?

6. *Absolutes versus probabilities in ethical decisions.* Probably the most difficult yet crucial ethical point to be discussed is the one of relativism. When can one rely on the *probabilities* of an occurrence in reaching an ethical decision rather than await the definitive event, no

matter how unlikely. Many ethicists hold to beliefs that are at least stated in absolute terms. By contrast, an increasing number of medical problems tend to be decided in relative terms. Examples are implicit in the decision not to resuscitate as well as in situations when one employs such concepts as "the odds of recovering from certain types of illness with certain types of treatments," "cost–benefit ratio," "risk–benefit ratio," and "decision trees."

We have no illusions that we will *resolve* many or perhaps any of the above antinomial propositions with the treatments of scientific and ethical issues in this book, but they serve as the backdrop for more specific discussions.

Laboratory Neurobiology

From the point of view of a laboratory scientist who works with animals, months or years can be spent generating experimental results which are novel and reliable. Several more years, usually including the work of several laboratories, can test these findings for their generality across organs, species, and so forth. These steps can be especially difficult when dealing with mechanisms by which nerve cells generate a behavioral response. The usual instinct of such a scientist may be to hope that over a span of years, in pharmaceutical laboratories, surgical suites, etc., such findings can eventually be applied for the well-being of humans. Therefore it is sobering to reflect on the ethical questions involved in the application of any medical treatment and especially to realize that some problems may be particularly difficult in the treatment of the nervous system or behavior. There may even be some paradoxes.

Among the most vexing problems, those connected with diagnosis may be more difficult in psychiatric or neurological settings because of the complexity of the systems involved. Consider circumstances in which a proposed treatment is nearly a step function: Some patients may receive no treatment, whereas others get a treatment which is quite severe, for example, institutionalization or psychosurgery. *The abstract problem is the risk of imposing a step function on an underlying continuum.* Of course, where the distribution of severity of symptoms is really continuous, nothing can be done in just this framework. But surely there are circumstances where the difficulties of decision rest not on the nature of the disease but on the lack of certainty in diagnosis. So, in the most difficult cases of psychiatric diagnosis, for example, the continuous distribution of results actually comes from the probability as well as from the severity of a particular disorder. Moreover, in the complicated diagnoses which can be encountered in possible disorders of the nervous system or behavior, the lack of precision may result from inadequate conceptualization of a disease entity from the very start. That is, imperfect understanding of any organ and its functions, as we would admit exists with the mammalian central nervous system and human behavior, may trick us into assigning single names for groups of distinct disorders whose separate frequencies of onset, unable to be analyzed individually, give the appearance of a sloppily rising curve of severity. In the next chapter, Dr. Jean Endicott, involved in the generation of the *American Psychiatric Associations Diagnostic and Statistical Manual* will discuss problems of psychiatric diagnosis.

A second major problem: informed consent. The acceptance of a statement of informed consent involves the linguistic output of a potential patient. The doctor's understanding must be that the patient is using the language in the manner of the population to which the doctor belongs, an essentially statistical idea. But, broad areas of the forebrain can influence language use and in neurological or psychiatric cases these may be statistically abnormal. Stating the paradox crudely: *What is the meaning of informed consent in cases where the organ of informed consent is statistically suspect?* Dr. Macklin's first chapter deals with problems of informed consent, and her second discusses a related problem: treatment refusals by psychiatric patients.

Decisions in cases of severe neurologic damage pose other sets of problems involving the law and questions of distribution of resources, at least. Dr. Levy's chapter shows how the systematic analysis of quantitative data from coma patients at the Cornell Department of Neurology can lay the foundation for decisions which minimize risks as a society would define them. From a point of view tempered by both medical and legal practice, Dr. Beresford evaluates legal strategies for handling difficult neurologic cases.

There are no easy answers to the particular ethical questions treated in this book. In a multitude of cases late 20th-century developments of neuroscience and behavioral science, and the consequent advances in means of therapy, have led to good conceptualizations, diagnoses, treatments, and good results. Distinct from these, there are cases where our advances in techniques have not helped or have actually raised the ethical problem. One wonders which aspects of these ethical problems are theoretically irreducible.

If diseases of the nervous system and behavior pose unique ethical problems because of the special relation of the brain to human values and language, studies of the nervous system and behavior may offer a *corresponding* set of insights, eventually, about the origins of values. In other words, for a set of special ethical problems, neuroscientists may recognize a *complementary* set of opportunities to understand the origins of values as they appear in ethical systems. In the usual analysis of why a person would perform a particular behavioral response and whether it was right or wrong, one goes through a long sequence of steps involving logic and the calculations of probabilities, but is still left with matters of values which are "just there," without an easily understood rationale. The increasing explanatory powers of the neural and behavioral sciences may eventually offer for the first time the opportunity for a *closed logical system rationalizing choices of primary values*. Naturalistic ethics, a field of ethical philosophy, deals with this possibility but is poorly elaborated.

"We are the ethical creatures we are *because* . . . " One approach to completing that sentence uses evolutionary thought. In its broadest form, this approach says that it has been adaptive for particular groups of animals to behave in particular ways toward each other. Occurrences of altruism and restrictions on the intensity of intraspecific aggressive behavior, for example, lead to animal populations where the young have a greater chance of survival to the age of reproduction. Discussing these biologic observations and their implications brings us to the area often called *sociobiology*, which we attempt to discuss without the polemics that surround that field. Dr. Arthur Caplan's chapter treats the implications of sociobiological findings for ethical thought. Dr. Jerram Brown expresses an understanding from evolutionary theory of how altruistic behavior comes about, and generalizes to a biological perspective on cooperative and competitive social behaviors. Dr. Colin Beer's discussion of intentionality, again, tends to favor extrapolating from animal behavior to concepts of human psychology.

If one approach to naturalistic ethics is from evolutionary thought, the other way of constructing a closed logical system for understanding why we have the values we do comes from the possibility of understanding neurophysiologic mechanisms at the very base of a human's behavior when he makes an ethical decision which manifests an understanding of right and wrong. That sounds very difficult, but understanding any vertebrate behavior is difficult, and the order of difficulty may not always be what is expected, *a priori*. For example, we know embarrassingly little about certain reflexes in the cat spinal cord but are making surprising progress in understanding an entire circuit for certain behaviors. The final chapter argues that understanding the physiology of the exquisitely ethical principle of reciprocal altruism may actually be no more difficult than many of the behaviors we try to explain as neuroscientists.

In no sense is this book "complete." Instead, it is intended to include informed discussions of the most difficult theoretical problems and statements of the theoretical opportunities encountered at the interface between ethics and modern neurobiology.

References

1. Hunt, R., and Arras J., (eds.) (1977). *Ethical Issues in Modern Medicine.* Mayfield, Palo Alto, California.
2. Ramsey, P. (1970). *The Patient As Person.* Yale University Press, New Haven, Connecticut.
3. Reiser, S. J., Dyck A. J., and Curran, W. J., (eds.) (1977). *Ethics in Medicine.* MIT Press, Cambridge, Massachusetts.
4. Weir, R. F., (ed.) (1977). *Ethical Issues in Death and Dying.* Columbia University Press, New York.

Part I

Special Ethical Problems in Dealing with Neural and Behavioral Disorders

Chapter 2

Making and Using Psychiatric Diagnoses: Ethical Issues

JEAN ENDICOTT, PH.D.*

Publication of the Third Edition of the *American Psychiatric Association's Diagnostic and Statistical Manual* (DSM-III) has brought some of the less obvious ethical issues involved in the clinical, administrative, and research use of diagnoses of mental disorders into sharper focus.

DSM-III is the result of years of work by many members of the mental health profession. Those who worked on it attempted to develop a system that would better serve the purposes of psychiatric diagnosis. The major purpose of classification of any set of medical disorders is to identify those conditions which, because of their potential negative consequences, implicitly have a call to action to the profession and to society. The call to action to those in the mental health profession is to (1) provide treatment; (2) to attempt to eliminate or reduce harm to the patient and to others; (3) to try to prevent the development or recurrence of the condition; and (4) to conduct appropriate research. The call to society also takes many forms, including various exemptions from certain responsibilities, provision of means for delivery of care, an attempt to limit any harmful consequences of the condition on others, and meeting the needs of patients with the condition, of their families, and of those who interact with them.

As in other areas of medicine, the immediate purpose of the

*Department of Psychiatry, College of Physicians and Surgeons, Columbia University, New York, New York 10032.

classification of mental disorders is to enable health professionals to communicate with each other about the subject matter of their concern. Ideally the use of a specific diagnostic label carries with it a "guarantee" that a careful differential diagnostic evaluation was made. Such a diagnosis often implies knowledge of the most likely prognosis for certain outcomes, response to various treatments, and possible psychobiological processes involved in establishing or maintaining the condition.

Making and using psychiatric diagnoses involves actions or inactions on the part of individuals that have the potential to either help or harm the patient as well as others. Therefore, many obvious as well as hidden ethical issues are involved in these actions.

Deliberate Use of False Diagnoses

Of course false diagnoses of mental disorder have been deliberately used by some for nonmedical purposes to control or harm others (e.g., with political dissidents, in custody suits, to gain control of wealth). Such calculated use of false psychiatric diagnoses to achieve non-medical aims is clearly unethical. There is no way a diagnostic system, good or bad, can directly affect such use since both the "information" on which the diagnoses are based and the diagnoses themselves are deliberately falsified. Although the use of psychiatric diagnoses to harm others is a continuing problem within the profession, it will not be discussed further in this chapter.

There are some other areas in which the deliberate misuse of diagnostic labels often takes place (e.g., when clients wish to qualify for third-party payment for psychotherapy or disability payments, to avoid or get out of military service). Here, as elsewhere, the degree to which one considers the behavior of the diagnostician unethical depends upon one's attitude toward "absolutes" of honesty as well as one's weighing of the potential consequences, good or bad, for the individual and for society as a whole. Since such cases also represent deliberate misdiagnosis, they will not be discussed further in this chapter either.

It is in those circumstances where a clinician is clearly trying to use psychiatric diagnoses for legitimate medical purposes that ethical problems arise because of the potential consequences of alternative actions. These everyday ethical problems of diagnosticians are the focus of this chapter.

Performing the Differential Diagnostic Assessment

Most individuals who are responsible for making psychiatric diagnoses like to operate as if the only problems involved are those associated with the need for careful evaluation, consideration of diagnostic criteria, and making differential diagnostic judgments (taking into account the likelihood of error and giving due respect to the need to continuously reevaluate the diagnosis once it is made).

Unfortunately, ethical issues are raised immediately, even in the initial process of diagnostic evaluation. DSM-III contains specific inclusion and exclusion criteria for making diagnoses which are, for the most part, descriptive. They do not depend upon the use of any particular set of theoretical or other explanatory principles. Furthermore, whenever possible, the criteria are based upon the results of research studies, particularly those relating to the importance of different clinical features in the prediction of genetic or prognostic differences (including response to different treatments, likelihood of recurrence, etc.).

The inclusion of such criteria is the major way in which DSM-III differs from DSM-II, the previous manual, or the current edition of the International Classification of Diseases. In contrast to DSM-III, they contain relatively vague descriptions of a smaller number of conditions and devote little attention to differential diagnosis. A consequence of the lack of carefully defined criteria of systems designed for routine clinical use has been poor reliability and a lack of clear-cut prognostic or genetic correlates. Furthermore, differential diagnosis has not been highly valued by mental health professionals in the past. It has been common knowledge that clinicians did not take great care and used greatly varying criteria when making diagnoses. Therefore, diagnostic evaluations made by others were not taken very seriously by sophisticated clinicians, administrators, or investigators. This was not as bad a problem as one might think because there are few known correlates for the diagnoses.

Presumably this will change with the widespread use of DSM-III, since there are specific rules for the diagnoses. Furthermore, many efforts are underway in this country to train clinicians to use it properly. It has generally been accepted as vastly superior to DSM-II, and superior reliability of the categories was demonstrated in field trials and reported in the manual. In the future, when a clinician uses a DSM-III category, the assumption will be made (by most of his colleagues) that he has carefully considered the criteria, obtained adequate information to apply them, and has addressed the differential diagnostic issues adequately.

There would be little disagreement that the deliberate failure to apply the criteria correctly, with the possible consequence of misdiagnosis in mind, would be considered unethical by most people. However, even when a clinician is trying to do the evaluation, he may not be able to do so. Diagnoses are often made in settings which almost preclude the kind of careful consideration that a good differential diagnosis by DSM-III criteria requires. Many clinicians are expected to make diagnoses when the available information is grossly inadequate to do the task well. This may be due to circumstances such as a large patient load that reduces the amount of time that can be spent with the patient (due to lack of resources or to greed), to reluctance on the part of the agency to obtain old records or seek other informants, to reluctance of the patient to be candid (e.g., prison, military), or to the limitations of the clinician himself in his ability to ascertain the needed information (e.g., language barriers, lack of knowledge of the criteria, lack of knowledge of cultural differences that affect their application.

In recognition of such circumstances (i.e., those times when it is unreasonable to expect a clinician to make a reasonable "final" diagnosis), DSM-III has categories whereby uncertainty may be indicated. These include the categories of "diagnosis deferred," "unspecified mental disorder (nonpsychotic)," and "atypical psychosis." One of these diagnoses should be used unless a clinician is satisfied that he has used all sources of information of potential relevance to the diagnostic evaluation (including laboratory tests when appropriate), and that the patient clearly meets criteria for one of the specified disorders listed.

To the degree that a clinician continues to work in a setting which interferes with the adequacy of diagnostic evaluation and continues to make DSM-III diagnoses without indicating their provisional nature, ethical issues are apparent (although they would probably be denied or unrecognized by many who do so). The potential negative consequences of misdiagnosis would lead some clinicians to choose alternative actions such as (1) to refuse to make diagnoses under such conditions; (2) to attempt to change the conditions directly or indirectly (such as alerting authorities); or (3) to clearly indicate that the diagnoses are made under less than ideal circumstances, with limited information, are provisional at best, and to indicate those differential diagnostic issues that could not be addressed adequately.

Keeping Aware of the Potential Consequences of Misdiagnosis

Some might ask: "What difference does it make if the diagnosis is wrong?" Such a question would rarely be asked for other medical diagnoses, but still arises at times from uninformed mental health professionals.

At the time of the development of DSM-II (1966–1969) little was known about the differential efficacy of various treatment modalities for specific mental disorders. This is no longer the case. There are considerable data available relating to some of the disorders listed in DSM-III where the potential consequences of receipt of the wrong treatment as well as the failure to receive the appropriate treatment are well established. Furthermore, the possible immediate as well as the long-term effects upon the patient, his family, as well as society, are also fairly well established for some DSM-III conditions. These may affect not only treatment choice but also other actions such as long-term institutionalization, family planning, etc.

A single example will be given here. If a patient with chronic DSM-III Bipolar Disorder (manic-depressive disorder) is given the misdiagnosis of DSM-III Schizophrenia, there are a number of possible negative consequences. He would probably be treated with a neuroleptic (which is likely to be ineffective and possibly countertherapeutic) rather than lithium and/or an antidepressant or ECT which would have a higher probability of being effective. His risk for suicide may be under-estimated as is his risk for the development of tardive dyskinesia. He is more apt to spend many years in institutions with a chronic cycling, recurrent, or "burned out" clinical picture. His level of social impair-ment would obviously be markedly affected. Family members and others (including mental health professionals) are more likely to be pessimistic and to assume that the patient will be chronically and severely impaired for the remainder of his life. On the other hand, a correct diagnosis of Bipolar Disorder, followed by appropriate treat-ment, has the potential of resulting in full recovery without significant relapse.

Mental health professionals who fail to make efforts to remain aware of the possible short- and long-term consequences of misdiagnosis, of newly available treatments for specific conditions, and of the other correlates of specific diagnoses are derelict. To the degree that they deliberately fail to do so and continue to represent themselves as experts and to make treatment recommendations on the basis of their diagnoses, their behavior would be viewed as unethical by many.

Assuring Appropriate Treatment for Patients

Assuming that the correct diagnosis has been made, it is the respon-sibility of those in contact with the patient to try to assure that he receives the most appropriate treatment.

Therapists are expected to be familiar with the relative efficacy of alternative treatment modalities for different conditions (or to seek such information when confronted with a diagnostic condition where they do not have that knowledge). If they know what is most likely to be effective and do not or cannot offer it themselves, they would generally be expected to share this information with the patient and/or his/her family and to make an appropriate referral. Unfortunately this is often not done. To the degree to which a deliberate failure to seek the most appropriate treatment increases the risk to the patient (e.g., of suicide, business failure, prolongation of a painful and socially disruptive disorder) and results in the expenditure of funds on less effective or ineffective treatment, it is unethical. Some clinicians admit that their concern over the loss of income reduces the frequency of such referrals, while others deny the potential harm to the patient or rationalize their failure to make appropriate referral in other ways ("he wouldn't accept it anyway").

Unfortunately, many clinicians protect themselves from being faced with the need for referral by remaining ignorant of the therapeutic implications of certain conditions, particularly if they are restricted from their use (e.g., psychologists and social workers may treat depressed patients but not inform themselves of the indications for somatic therapy of such patients). This will be an increasingly important ethical issue for mental health professionals as the number of studies indicating the relative efficacy of various somatic treatments for certain DSM-III conditions accumulates. In time, the legal consequences of the failure to seek more effective treatment (i.e., malpractice suits) may increase the number of appropriate referrals even if recognition of the ethical issues involved does not.

Other clinicians who are not directly responsible for the treatment of patients often become aware of either misdiagnosis or lack of appropriate treatment. This often happens in the course of research studies where independent diagnostic evaluations are made and treatment data are collected. Some investigators choose not to share this information, while others feel ethically required to tell the therapists (and if that does not appear to result in action), the patient and his/her family. Often no "policy" decision is made by the investigator regarding such instances until a particularly dramatic situation arises. A failure on the part of an investigator to directly address the conflicting needs of the study design, desire for cooperation from the therapists, etc., with the needs of the individual patients, often leads to conflict among staff over the ethical issues involved. Although there is no single "right" decision, the possible consequences of the withholding of the information against other needs should be weighed in each case. Deciding to ignore the issue is in itself an action which presents ethical problems.

Sharing the Diagnostic Information

Once a diagnosis is made and the clinician is aware of the correlates of such a diagnosis, he/she is often faced with a problem, particularly if there is a high likelihood of instability, impulsive behavior, breaks in reality testing, violence, recurrence of the disorder, or chronicity with deterioration, etc. There are a number of "classes" of people other than the patient for whom such information is of great potential importance—the family, present or potential employers, the people in the community in which he lives, legal authorities. While obvious conflicts between the needs of the patient for privacy and a right to earn a livelihood in the job of his/her choice and those of society rarely arise as dramatically as is the case with contagious diseases (e.g., active T.B., some venereal diseases), they do arise fairly frequently.

Who is to be given diagnostic information regarding the patient? The patient (he/she may lose hope)? His/her family (they may refuse to take the patient back)? His/her fiance (the wedding may be cancelled)? His/her boss or potential employer? The police? Federal authorities (risk to the president)? Insurance companies? This question sometimes is viewed as irrelevant and self-evident by private practitioners, but even they are at times faced with the conflicting needs of the individual patient, his/her family, and society.

Clinicians who work in institutional settings are more frequently faced with such decisions. Certainly in a military, school, or business setting, knowledge of certain diagnostic conditions almost always has the potential for conflict between the needs of the patient and those of the institution.

Sharing of the diagnostic information depends, in part, upon the "contractual" relationship between the diagnostician, the patient, and the person or agency calling for or paying for the evaluation. In addition, the potential consequences of sharing or not sharing the information with specific "others" are related to the nature of the ethical problems encountered.

That there are often negative reactions to the knowledge of a diagnosis of mental disorders in others is well recognized. How much of the reaction is based upon misinformation and fear and how much on the real correlates of certain conditions varies. This must obviously be taken into consideration. For example, the clinician may have to weigh the ethical considerations regarding the patient's need for a job versus society's need for the stable and safe performance of certain tasks. The degree to which the condition of the patient is irrelevant to the requirement of the job makes the decision easier, but one would want clinicians to weigh potential consequences when such decisions are made (e.g., should a person with recurrent manic episodes refractory to lithium treatment fly for a major airline, or have ready access to and be in charge of large amounts of money?). Such questions produce conflicts in diagnosticians and among clinical staff at times. Unfortunately, there are no general applicable guidelines to indicate where the greater need lies, although at times it is fairly apparent. The issues involved are no different from those that arise in other fields of medicine (e.g., should a patient with hypertension, premature ventricular contractions, and a history of a past M.I. be a pilot for an airtaxi service?).

Some clinicians attempt to ignore the ethical issues altogether regardless of the possible consequences of sharing or not sharing the information. On the one hand, there are those who hold that confidentiality is the only value, and on the other, that those who pay for the evaluation or contract for the evaluation deserve to receive all information. Hiding behind "policy" does not change the fact that the two kinds of clinicians are making decisions which will raise ethical issues at times.

The developers of DSM-III tried to provide information relevant to the likelihood of recurrence, the most common complications of the disorders, and the short- and long-term prognosis of those conditions where such knowledge was available. The intention was to decrease the negative expectations regarding mental disorders in general while acknowledging that the prognostic expectations for some conditions are relatively poor. Unfortunately, the point was not made as strongly as it could have been that the predictive value of the information for the individual patient is low and that past history of behavior is often a more powerful predictor than is the diagnosis of the current condition.

The careful and deliberate sharing of diagnostic information coupled with an attempt to educate those receiving the information as to the short- and long-term prognosis (including potential effects on specific kinds of work, etc.) would appear to be useful in some circumstances. Unfortunately, this is often impossible.

It is well known that when diagnoses are made available without explanation to school systems, businesses, or insurance companies, the patient's interest may be compromised. Since most clinicians are aware that psychiatric diagnoses submitted for third-party payment often find their way into nonmedical records and may be used for nonmedical decisions, they often deliberately give the "least potentially damaging" diagnosis that will still qualify for coverage. A consequence of this is that no one thinks of such data as being "real data." In a similar fashion, the diagnoses listed on examinations for life insurance are often highly suspect and thought to represent an under-reporting of more serious conditions.

DSM-III reflects the developers' awareness of such issues. Names of conditions were selected with an effort to eliminate those with obvious negative associations (e.g., hypochondriac, psychopath, hysteric), and to limit the use of terms traditionally associated with assumption of chronicity to those conditions where there is a very high likelihood of chronicity or where it is part of the definition itself (e.g., schizophrenia, dysthymic disorder). Terms or principles which carried implications of negative mechanisms were also dropped when possible (e.g., conversion reactions). An effort was made to use descriptive terms which would not arouse immediate negative reactions in those hearing them.

However, to the degree that DSM-III reflects current knowledge about the likelihood of recurrence, chronicity, poor social and work performance, and other complications or risk of inheritance of the specific disorder, the information is of obvious relevance for all concerned. To ignore these correlates, when considering the "need to know" of specific parties, will obviously raise ethical issues.

Unfortunately, the "need to know" issue is also undergoing a process whereby there is establishment of "guidelines" through the courts. There have already been instances where therapists with knowledge of relevance to judgments of potential risk for suicide or violence have been sued for failure to share the information. It is probably only a matter of time before such suits are brought about on the basis of knowledge of specific psychiatric diagnoses and specific jobs or responsibilities on the part of the patient, and failure on the part of the clinician to share the diagnostic information.

Keeping Oneself Aware of the Use and Abuse of Psychiatric Diagnoses

Some clinicians deliberately, or unconsciously, keep themselves unaware of the use to which others put psychiatric diagnoses. This is constantly changing, and has obvious effects upon individual patients and others (e.g., coverage for inpatient or outpatient care, disability payments, life insurance coverage, likelihood of continued institutionalization, likelihood of acceptance in specific training programs, sheltered homes, jobs). To act as if there were no individual or social consequences to the use of diagnostic information provided is to ignore the ethical considerations involved.

Reporting of Research Findings

Often research investigators act as if their studies and their reports have no potential for harm (particularly if they are not involved in treatment studies). This is not the case. Even studies which focus upon diagnostic issues, such as comparing sets of criteria or seeking information on their correlates, can be conducted or reported in such a fashion that there is potential for harm to the subjects themselves or to others who have the specific disorder being studied.

There is no question that some investigators behave in an unethical fashion (e.g., break promises of confidentiality, use case vignettes which allow the easy identification of particular subjects, or collect the data in a way that is embarrasing to the subject or risks his relationships with his co-workers or family). It is easy to identify and deplore such actions.

However, many investigators seem unaware of potential harm or other ethical issues when they report the results of their studies. They can raise false hopes with sensationalistic or exaggerated claims (even in such areas as "finding a genetic marker" for affective disorder). They can increase the feelings of hopelessness in those who have chronic disorders, or negative reactions in those in positions to hire or work with them. They can make statements that raise the risk for embarrassment or harassment of patients with the condition.

Given the increasing trend for newspapers and magazines to write follow-up articles based upon scientific reports and to modify the stress given to particular aspects of the findings, investigators must become increasingly aware of the potential use to which some of their statements may be put. They should ask themselves about the potential harm if certain statements are quoted out of context. They should make it clear that they are speaking of findings for groups of patients that may or may not apply to individual patients. They should not act as if they are "just reporting" the facts and have no responsibility for the consequences. At the same time, they must report all relevant facts, negative as well as positive.

As mentioned previously, sometimes an investigator may be faced with a situation in which he/she must choose to act or not act in the patient's behalf. He/she must weigh the potential consequences for everyone involved, including the research staff and the funding agencies. Here as elsewhere, such choices are difficult, yet the need to make them does arise, and acting as if there were no ethical issues, or being unaware of the ethical issues, has potential for harm as well.

Conclusions

There is no way to estimate the true magnitude of the intended or unintended deleterious effects of the use of psychiatric diagnoses.

No diagnostic system, no matter how reliable or valid for various specific purposes, can assure that the diagnoses will be made and used in a fashion considered to be ethical by all concerned. This is true for the diagnostic systems for all medical disorders. The nature of medical conditions (i.e., their potential for negative short- and long-term effects on the patient, those closely associated with him/her, and even society) will inevitably lead to conflicts in the needs of those involved.

At times, there appears to be more controversy regarding the criteria for psychiatric disorders, their application, and their use than for other medical conditions. This undoubtedly arises from many sources. At the same time, some diagnoses share with some other medical conditions an almost automatic negative reaction on the part of those who obtain the information (e.g., venereal diseases, cancer, epilepsy). The potential harmful effects of psychiatric diagnosis are probably exaggerated (perhaps in attempts to use psychiatry as a scapegoat for societal failure to deal with difficult social problems or the painful reality of disability that is intrinsic to some chronic psychiatric disorders).

Unfortunately the use of the new DSM-III has the potential to increase as well as decrease the negative consequences of the use of psychiatric diagnosis. To the degree that it improves diagnosis for conditions that can be successfully treated, it may decrease the assumptions of chronicity on the part of others and decrease the fear and dread associated with certain conditions. To the degree that it increases knowledge regarding the likelihood of chronicity or re-currence, of medical complications, or of risk for passing it on to one's children, it has the potential for increasing the risk of harm to individual patients. Since such data will be true (on the average) for groups of patients with that condition and may not be true at all for the individual involved, it makes the choice of actions to be taken more difficult.

In the past, psychiatric diagnoses were made using vague and unreliable criteria for conditions with few known correlates. There were fewer obvious negative consequences of wrong diagnoses since there were fewer effective treatments available. Since less was known about the conditions, there was also less reason to share information with others. It is almost inevitable that the number of situations which call for action on the part of the diagnostician or clinician will increase and as that increases, so will the number of conflicts associated with ethical issues. This being the case, those who make and use psychiatric diagnoses should become more aware of the ethical issues involved and make their decisions regarding alternative actions after weighing the potential consequences, rather than in a blind fashion as if there were no potential consequences or choices open to them. They should not take their actions for granted or be surprised when they are questioned or criticized because of them.

Chapter 3

Problems of Informed Consent with the Cognitively Impaired

RUTH MACKLIN*

I

The doctrine of informed consent for treatment and research is by now firmly embedded in health law and medical ethics. Federal regulations govern all research conducted on human subjects and supported by federal funds (Code of Federal Regulations, 1981); many states have passed legislation that mandates informed consent for treatment (Meisel and Kabnick, 1980); and the common law contains an increasing number of cases dealing both with informed consent for therapy and for research. Although it is interesting to learn the philosophical bases for the doctrine of informed consent (Veatch, 1978; Donagan, 1977) and to trace its history in law in the United States (Simpson, 1981; Trichter and Lewis, 1981), this chapter will be devoted primarily to the concept of informed consent as a moral requirement in the biomedical domain.

Difficulties surrounding implementation of that moral requirement are not confined to patients who suffer some cognitive impairment. Indeed, those who doubt the possibility of ever—or, at least, usually—obtaining fully informed, wholly voluntary consent from patients for purposes of treatment or research express their skeptical concerns even with regard to ordinary patients and normal, healthy research

*Department of Community Health, Albert Einstein College of Medicine, Bronx, New York 10461.

subjects. Before approaching the problems of informed consent that arise with patients of questionable mental capacity, it would be useful to review briefly the kinds of doubts typically raised regarding persons presumed to be mentally normal. Since the difficulties surrounding patients of questionable capacity can be treated as extensions of problems claimed to exist in cases of persons who are physically but not mentally ill, a brief look at the main skeptical challenges will set the stage for the later discussion.

The key features of informed consent are, first, the process of informing the patient or subject: How much information, of what specific kinds, and in just what form, must be conveyed in this process? This will be referred to as the element of *disclosure*. The second feature is the patient's or subject's understanding of the information conveyed. This element can be broken down further into several additional components: understanding at the time the information is conveyed; recall of that information at some later time; and the individual's appreciation of the significance of the information. For now, let us refer to this element simply as *understanding*. The third element is the *voluntariness* with which consent is granted. Federal regulations stipulate that investigators "shall seek such consent only under circumstances . . . that minimize the possibility of coercion or undue influence" (45 CFR, 1981; 46.116). In treatment contexts, physicians, all too familiar with how easily their patients may be swayed by a healer in the role of expert and authority figure, often proclaim that they can get their own patients to consent to virtually anything. When we reflect on the facts that patients are often beset by fear and anxiety surrounding their illness or its prognosis, that they sometimes express unwillingness to hear information concerning the risks or side effects of medically indicated procedures, and that they are often willing or even eager to delegate decision-making authority to their physician (Meisel and Roth, 1981; Cross and Churchill, 1982), it is evident that all three elements of informed consent may be fulfilled only in small part in a range of typical encounters in the biomedical setting.

By way of further introduction, it is worth remarking on the moral purposes behind the doctrine of informed consent—purposes that can be articulated apart from the legal sources and the causal factors that led to the adoption of informed consent requirements. There appear to be two distinct yet related purposes that informed consent is meant to serve. Each stems from a different prominent tradition in Western moral philosophy.

One tradition, utilitarian moral theory, underlies the aim of protecting patients and research subjects from potential harm. The use of risk–benefit equations, in which a proposed treatment or research protocol, in order to be ethically acceptable, must promise the likelihood that the benefits outweigh the risks, is an expression of the principle of utility. Otherwise known as the "greatest happiness principle," that ethical precept holds that right actions or practices are those that result in a balance of pleasure over pain, happiness over unhappiness, or other beneficial consequences over undesirable ones, for all persons who stand to be affected (Bentham, 1789; Mill, 1863). Utilitarianism is known as a *consequentialist* theory of morality, since it takes the consequences or results of actions to be the morally relevant factors for making ethical judgments. Risk–benefit and cost–benefit analyses are straightforward applications of the utilitarian moral principle.

The utilitarian principle plays another role in justifying the actions of physicians and biomedical researchers, in addition to that of underlying risk–benefit analyses. Put in another form, the principle of utility can be understood as a precept of beneficence (*Belmont Report*, 1978), stating that one should do good whenever possible, or at least, that one should minimize harms. That precept can be used to justify research projects that may place experimental subjects at risk with no benefit to them, but with the promise of producing overall good to future patients, which is a variant of its application in risk–benefit equations.

In a quite different context, however, it can be used to justify paternalistic behavior on the part of physicians toward patients. The term "paternalism" is understood here to mean "the interference with a person's liberty of action justified by reasons referring exclusively to the welfare, good, happiness, needs, interests or values of the person being coerced" (Dworkin, 1972). It is important to note that it is the coercer's *perception* of what is for the good or benefit of those others, and not the latter's perception, that characterizes paternalistic behavior. Later in this chapter and again in the next, paternalism will reappear. The point of introducing the notion here is to illustrate how the principle of beneficence, in one form or another, is one of the moral values that gives rise to dilemmas when two or more values are in conflict and cannot simultaneously be satisfied.

Utilitarianism is not the only major philosophical theory that underlies ethical principles used to justify actions or practices in the biomedical domain. The other prominent theory, equally embedded in the Western moral and legal tradition, is that of Immanuel Kant. This ethical perspective emphasizes the rights and duties of persons, rather than the consequences of actions. In the specific form put forward by Kant, the morally relevant charateristic of right actions is the *motive* from which they are done. For Kant, the only motive that can properly justify an action is that of *duty*. An action must not merely accord with what duty prescribes; it must be performed out of a sense of duty, in order to be truly moral (Kant, 1785). How we come to know our duties

is an epistemological problem in Kant's theory and that of others who
adopt this moral perspective, known as deontology. Inquiry into the
epistemological features of deontological theories is beyond the scope
of this discussion, however important and interesting it may be for a
thorough philosophical examination. For our purposes here, it is
sufficient to note the differences between the utilitarian (or consequentialist) approach and a deontological theory, as those two per-
spectives relate to the moral foundations of informed consent.

Often referred to in the field of bioethics as "respect for persons"
(*Belmont Report*, 1978), the ethical precept that underlies the Kantian
approach is expressed in terms of the autonomy of the individual and
respect for human dignity. Typical of the use of this perspective in
bioethical contexts are claims couched in the language or rights: the
"right to life"; the right to refuse treatment; the right of patients to
decide on matters concerning their treatment; the right of a woman to
control her own body; the "right to die"; and a range of other alleged
rights that have been claimed on behalf of patients and research
subjects in biomedicine. Captured ironically in the title or content of
more than one article deploring recent court decisions that enable
mental patients to refuse psychotropic drugs, such patients have been
described as "rotting with their rights on" (Appelbaum and Gutheil,
1980a, 1980b, 1980c). In sum, the deontological perspective takes as
the central moral notions the rights and duties of persons, where the
concept of rights and duties are treated as correlative.

The point of describing these moral principles and the ethical
theories from which they arise are not simply to note that they explicate the dual purpose behind the informed consent requirements—
protecting people from harm and showing respect for persons. A
further reason is to articulate the potential conflict between these two
ethical precepts, a conflict of values that will emerge a bit later in this
chapter and again in the subsequent chapter on treatment refusals.
Since moral dilemmas in medicine, in particular, and in other applied
contexts generally, can often be traced to conflicts of values or clashes
between moral principles, it is useful to identify what those principles
are and to see why the dilemmas are so intractable.

The ethical principles just described are easier to state and to
explicate than they are to apply. In the case of utilitarianism, an
essential ingredient in such applications is an accurate assessment of
the facts and probabilities involved. Utilitarianism is a straightforward-
ly empirical theory of morality. It requires a good deal of inductive
evidence, drawn from an individual's own experience, from the
reported experiences of others, and from wisdom gleaned from the
whole of human history. In order to apply the utilitarian principle
correctly, one must have some knowledge of the probabilities of

different alternative outcomes of actions, a basic awareness of what generally produces pleasure or happiness in people, what is likely to result in happiness or unhappiness for the particular persons involved, and a host of other facts relevant to assessing probabilities and predicting outcomes.

Deontological theories do not place a central reliance on facts or empirical data, but they surely require the ability accurately to assess the motives of actions (one's own and those of others), as well as to determine when individual autonomy is being violated and when it may be absent in the first place. Now the concept of autonomy is frequently invoked in discussions of informed consent, especially in instances where paternalism on the part of physicians is suspected or in which patients refuse to grant consent for treatment. But that concept is systematically ambiguous, and at least two clearly different meanings can be distinguished. The first of these is "autonomy as free action," (Miller, 1981), usually taken to be synonymous with the concept of self-determination. The second meaning, which is a much richer concept, can be roughly stated as "autonomy = authenticity + independence": "the autonomous person is one who does *his own* thing" (Dworkin, 1976). A fuller discussion of autonomy, explicating some additional meanings, appears in the following chapter in connection with treatment refusals. The importance of distinguishing among the different senses of the concept should be recognized at the outset, however, for the following reason. If a person appears to lack autonomy in the second of the two senses just noted, whether by virtue of illness, injury, organic brain syndrome, "brainwashing," or any other causal factor, the question arises whether there is any genuine autonomy to be violated in overriding treatment refusals or bypassing informed consent procedures. It is this issue—in part factual, in part conceptual, and in part ethical—that forms the central focus of problems of decision making involving the cognitively impaired.

II

It needs little reminder that there are no simple solutions to the problem of obtaining informed consent from cognitively impaired patients or research subjects. In addition to the central morally relevant factor already mentioned—that competing values or ethical principles are likely to be in tension in such cases, a number of additional factors contribute to the complexity of the issue. The following questions, grouped under relevant topical headings, serve to identify the different considerations that need to be taken into account.

Characteristics of the Patient or Subject

1. Is the patient clearly incapable of granting consent for treatment or research, or instead, of questionable mental capacity?
2. Is the cause of impairment known? If so,
    (a) Is the impairment reversible (as in cases of dementing illness caused by poor nutrition or drug interactions)?
    (b) Is the impairment the beginning of an inexorable or predictable decline (as in cases of senile dementia of the Alzheimer's type)?
    (c) Might an improved level of cognitive functioning be achieved by administering psychoactive drugs?
3. If the patient is mentally incompetent, is he or she one who has never been competent (because of mental retardation), or is the lack of present capacity a result of recent disease, accident, or injury?

Characteristics of the Situation

4. If recent illness or injury has impaired the patient's mental capacity, has there been a legal declaration of incompetence and appointment of a legally authorized guardian for the purpose of granting consent and participating in treatment decisions?
5. How much weight should be accorded the wishes of family members in cases where they are not legally empowered to grant consent on behalf of a patient who demonstrates cognitive impairment?
6. Is consent being sought for treatment or for research? If the latter,
    (a) Does the research stand to benefit the patient directly (as in cases of experimental drugs or innovative procedures)?
    (b) Is the research "non-therapeutic," that is, potentially benefitting persons other than the patient or research subject?

Setting the Standard for Determining Competency

7. If the patient is of uncertain mental capacity, where ought the standard of competency be set: very low, and thus most respectful of autonomy (in the sense that means "self-determination")? Or very high, thereby adhering to the value of beneficence (acting in the patient's "best interest" as medically determined)?
8. If administering two (or more) different tests typically used to assess competency yields different outcomes (for example, one of the

items on a mental status exam versus a standardized IQ test), which
should be determinative: the one that points to competency or the one
that suggests incompetency?

9. Should the particular biomedical procedure being proposed
influence the level at which the standard of competency is set? If so,
should it depend on

(a) The invasiveness of the procedure (brain biopsy, neurological
surgery)?

(b) The irreversibility of the procedure or its side effects
(neurological surgery, likelihood of tardive dyskinesia resulting from
phenothiazines)?

(c) The pain or discomfort of the procedure (electroconvulsive
therapy, side effects of psychoactive drugs)?

10. Should the level at which the standard of competency is set be
influenced by the probable consequences of *not* doing the procedure
(for example, possible suicide of a depressed patient, return of
schizophrenic symptoms in a patient refusing psychotropic drugs, swift
decline of a patient who refuses very risky brain surgery)?

11. Should the standard of competency be the same for assessing
(prospectively) the competency of a patient to grant informed consent
for treatment as it is for assessing the competency of a patient who has
already refused treatment?

Additional questions will no doubt arise as we reflect on the myriad
issues that come together when consent is sought from patients or
research subjects of uncertain mental capacity. The above list of
questions serves to identify the different factors that should be taken
into account regarding particular patients or subjects, the special
circumstances in which they are situated, and the types of uncertainties
that surround setting the standard of competency. The range of
relevant factors suggests that it may be desirable to set different
standards of competency in different contexts or for different purposes,
while still attempting to arrive at those standards systematically and
with reference to objectively determined criteria.

III

In the past several years, the issue of the competency of patients to
grant or refuse consent for treatment or research has received an
increasing amount of attention in the medical, legal, and philosophical
literature. The trend in both law and medicine is in the direction of
developing a notion of variable competence, that is, selecting situation-

specific criteria for judging competence, rather than viewing it as a global attribute of people. Among the leading researchers in this area are Alan Meisel, an attorney, and Loren Roth, a psychiatrist. Along with their colleagues, Roth and Meisel have undertaken numerous studies of the capacity of patients (especially psychiatric patients) to grant or refuse consent. The following account of their findings and analyses is drawn from two published articles, one devoted to tests of competency to consent to treatment (Roth, Meisel, and Lidz, 1977) and the second focusing on consent for participation in research (Appelbaum and Roth, 1982).

Appelbaum and Roth state that their own clinical experience, in additon to a review of the literature, shows that the various standards for competency that have been proposed cluster into four groups. These groups can be arranged in a hierarchy, such that each represents a stricter test of competency. The four standards of competency, from least rigorous to most stringent, are as follows:

1. *Evidencing a choice.* This test of competency requires that the subject of a research project actually communicate a decision about participating. This standard need not require the subject to verbalize that decision, however, since a behavioral criterion may be used: the subject's cooperation in the early procedures involved in a study. Because of the very low level of competency presupposed by this test, it is open to question whether or not subjects can properly be said to have "made a decision."

In the earlier article devoted to tests of competency to consent to treatment (rather than research), Roth, Meisel and Lidz (1977) state that "Under this test the competent patient is one who evidences a preference for or against treatment. This test focuses not on the quality of the patient's decision but on the presence or absence of a decision. . . . Only the patient who does not evidence a preference either verbally or through his or her behavior is considered incompetent." (Roth, Meisel, and Lidz, 1977, p. 280). According to these writers, "evidencing a choice" is a very low standard of competency and is the most respectful of the autonomy of patient decision making. They offer the following case example:

> A 41-year-old depressed woman was interviewed in the admission unit. She rarely answered yes or no to direct questions. Admission was proposed; she said and did nothing, but looked apprehensive. When asked about admission, she did not sign herself into the hospital, protest, or walk away. She was guided to the inpatient ward by her husband and her doctor after being given the opportunity to walk the other way. (Roth, Meisel, and Lidz, 1977, p. 280)

This case example not only demonstrates that the standard of competency is indeed very low but also raises doubts about the

meaningfulness of the author's reference to "the autonomy of patient
decision making." Even in the sense of "autonomy" that means "free
action" or "self-determination," there is considerable doubt whether
the patient in the example can be said to have *chosen* or *decided* to walk
to the inpatient ward. One writer highly critical of the inclusion of this
standard as a test of competency, asserts that "under this test, the only
incompetents are those who do not even temporarily share a world with
the rest of us: for example, the catatonic. . . . The ability to consent test
focuses entirely upon a result, and ignores the processes that gave rise
to that result." (Freedman 1981, p. 62).

2. *Factual Understanding of the Issues.*   Appelbaum and Roth (1982)
identify this next criterion as the single factor most widely accepted as
a standard for competency.

> A typical formulation requires that the subject have 'the cognitive
> capacity to consider the relevant issues.' Those areas that have been
> considered to be of crucial relevance for the subject to understand
> include: 'the nature of the procedure, its risks, and other relevant
> information,' 'the nature and likelihood of success of the proposed
> treatment and . . . of its risks and side-effects,' 'the available options,
> their advantages and disadvantages,' 'the knowledge that he has a choice
> to make,' 'who he is, where he is, what he is reading and what he is doing
> in signing the paper,' and 'the consequences of participation or non-
> participation.' . . . The rigor of the requirement of understanding ob-
> viously increases with the amount and complexity of material that is
> required to be understood. (Appelbaum and Roth, 1982, pp. 79–80)

According to the authors, "factual understanding" has been the
primary element of legal tests of contractual and testamentary
capacity. It can be interpreted in either of two ways: the requirement
that a patient or subject have the *ability* to understand; or the more
demanding criterion that the patient or subject manifest *actual*
understanding.

These two interpretations of the "factual understanding" standard
for competency are discussed separately by Roth, Meisel and Lidz
(1977) in their earlier article. Concerning "ability to understand," the
authors claim that "decision making need not be rational in either
process or outcome; unwise choices are permitted. . . . What matters in
this test is that the patient is able to comprehend the elements that are
presumed by law to be a part of treatment decision making. How the
patient weighs these elements, values them, or puts them together to
reach a decision is not important " (Roth, Meisel, and Lidz, 1977, p.
281). Among the elements the patients may be called on to show a
capacity to understand are the risks, benefits, and alternatives to
treatment. One potential problem contributing to uncertainty is the
possibility that a patient may understand the risks but not the benefits

or vice versa. A more bizarre possibility is that an individual may construe the risks as the benefits. The authors offer the following case illustration:

> A 49-year-old woman whose understanding of treatment was otherwise intact, when informed that there was a 1 in 3,000 chance of dying from ECT, replied, 'I hope I am the one.' (Roth, Meisel, and Lidz, 1977, p. 282)

The "actual understanding" test is potentially more difficult to attain, and it places a greater obligation on the physician to ensure that the patient has in fact understood. The authors note that the California law requiring the review of patient consent to electroconvulsive therapy implicitly adopts this test of competency to consent to treatment. Roth *et al.* view this test as more stringent than that requiring the ability to understand, claiming that it "delineates a potentially high level of competency, one that may be difficult to achieve" (Roth, Meisel, and Lidz, 1977, p. 282). Again critical of the tests, and especially of the alleged distinction between these two, Freedman (1981) writes:

> Practically speaking, there is nothing to recommend the first above the second; if we shall be testing his ability, we might as well inform him at the same time. . . . Since we are not going to second-guess the reasons which a person has used in making his choice . . . and since both the ability and actuality tests will be dealing with material of an equivalent complexity, if one is competent under one test he must be competent under the other as well. (Freedman, 1981, p. 62)

Freedman raises several objections against the "actual understanding" test, the most telling of which is a potential circularity. Asking how much information a person needs to understand to be competent, Freedman cites the trend toward specifying the amount of information "the average, reasonable, competent patient would require in order to make a decision. . . . To adopt the 'understanding' standard, we might have to engage in a hopeless, circular argument, defining competence in terms of understanding information, and defining the required information in terms of competence" (Freedman, 1981, p. 63).

   3. *Rational manipulation of information.* An even stricter standard of competency than demonstrating actual understanding is a patient's or subject's ability to use the information provided in the decision-making process. This standard falls under a heading variously termed "judgment," "rationality," "reality testing," and "decision-making capacity." Appelbaum and Roth (1982) note that this standard is embodied in legal rules concerning contractual and testimonial capacity, recognizing at least one defect of rationality—the presence of "insane delusions"—as grounds for invalidating an individual's acts

(Appelbaum and Roth, 1982, p. 82). However, since the concept of rationality is itself both vague and ambiguous, this test is likely to yield uncertain or conflicting assessments of competency.

4. Appreciation of the nature of the situation. This is the strictest standard of competency, and requires that "the rational manipulation of information take place in the context of the subject's appreciation of the nature of his situation" (Appelbaum and Roth, 1982, p. 84). The concept of appreciation as used in this test is analogous to that employed in the notion of criminal responsibility. There is an affective as well as a cognitive component in this sense of "appreciation," and the psychiatric tests of competency for this standard include "the usual techniques of judging insight and determining the presence of psychotic-level defenses" (Appelbaum and Roth, 1982, p. 87).

Noting that these four general categories yield multiple standards for competency of varying stringency, Appelbaum and Roth conclude that any of the four standards, or some combination of them, can be viewed as legitimate so long as they can be justified from some reasonable policy perspective (Appelbaum and Roth, 1982, pp. 87–88). Among the policy-oriented goals one may seek to attain are the following: *maximizing autonomy* (in the sense of self-determination of patient or research subject); *promoting rational decision making*; acting in accordance with a moral principle of *beneficence*; and exhibiting *respect for persons*. Although each of these goals is legitimate and may be appropriate to pursue under certain circumstances, taken together, they may be mutually incompatible. For example, a standard that requires no more than the patient's "evidencing a choice" would maximize that patient's autonomy in the minimalist interpretation of that concept, while achieving the goal of rationality in decision making would seem to require a very high standard for competency. And invoking the principle of beneficence is likely to conflict with the goal of maximizing a patient's autonomy, since "to the extent that the requirement for informed consent limits the freedom of the individual to consent to research under condition that may be acceptable to him, but not because risk is involved to society as a whole, it limits an individual's exercise of autonomy for the sake of his protection" (Appelbaum and Roth, 1982, p. 90). The same conclusions hold for psychiatric patients consenting to treatment: The stricter the standard of competence, in the interest of protecting patients from their own unwise decisions, the more autonomy is traded off for a gain in benevolent paternalism.

This discussion of various standards of competency is intended to encompass refusals to consent to treatment or research as well as patients' or research subjects' consent. Yet as has frequently been pointed out, there is a significant asymmetry in the medical profession's attitudes toward consent and consent refusals. A refusal to consent to treatment deemed medically necessary is often taken as evidence of a patient's lack of competency, even in the absence of a psychiatric evaluation. In a case that occurred in my own recent experience, an elderly woman suffering from heart block refused to consent to the insertion of a temporary pacemaker. A psychiatric consultation was sought, and the finding was that the patient was not incompetent to grant or to refuse consent by all the usual tests, and that she was not psychotic. Yet because of the patient's seeming failure to appreciate the probable consequences of her refusal, the psychiatrist held that although competent, the patient exhibited "denial of psychotic proportions." Roth, Meisel, and Lidz, (1977) sum up the situation in their discussion of the test of competency requiring that the patient make a choice having a "reasonable" outcome:

> Ultimately, because the test rests on the congruence between the patient's decision and that of a reasonable person or that of the physician, it is biased in favor of decisions to accept treatment, even when such decisions are made by people who are incapable of weighing the risks and benefits of treatment. In other words, if patients do not decide the 'wrong' way, the issue of competency will probably not arise. (Roth, Meisel, and Lidz, 1977, p. 281)

Yet the question remains whether the standard of competency should be the same when consent is sought initially and when competency may need to be reevaluated following a patient's refusal to consent. To move back and forth between two standards in this situation may well be to stack the cards in favor of the medical practitioner's decision rather than the patient's. Yet because of the bad consequences that may result from consent refusals by patients of questionable competence, a reluctance to set a low standard of competence for patient refusals reveals an understandable bias toward the principle of beneficence by medical practitioners. These last considerations will be addressed in greater detail in the following chapter.

However it is decided on which side it is better to err when dealing with patients of uncertain competence, it should be emphasized that different moral principles and value commitments underlie each of the proposed standards of competency. Proponents of a loose or weak test for competency most likely hold a strong commitment to individual liberty and to the priority of autonomy over competing values. Those who opt for stringent tests of competency more than likely support the legitimacy of benevolent paternalism in medicine, placing the health, well-being, and survival of patients above their freedom and autonomy. These clashes lie at the heart of the realm of disputed values, a realm in which reasonable people disagree on which values should take precedence when two or more widely held moral principles come into conflict. There is no wholly satisfactory way of resolving such ultimate value disputes, and in the absence of an agreed-upon method for arriving at a solution, the best approach any individual can take is to seek good reasons to justify the preferred solution.

IV

Whether a standard of competency is selected so as to be strict or lenient, variable or fixed, on the side of promoting autonomy or allowing for paternalism, there remains the problem of decision making on behalf of those judged incompetent. Strictly speaking, according to law, in order for anyone other than the patient to grant or refuse consent (except in cases of emergency), there must be a judicial determination of incompetence. Once that legal declaration is made, it falls to the next of kin or to a court-appointed guardian to consent for treatment or research on behalf of an incompetent person. In practice, however, these legal requirements are not always adhered to, and family members are often asked to grant consent, especially for elderly relatives, when cognitive impairment is evident or when a psychiatric consultation is sought and the patient is evaluated as having poor mental status.

Although law and morality intersect, and a presumption should exist in biomedical practice, as elsewhere, in favor of obedience to law, it is often instructive to consider what the law ought to be, in addition to knowing what it in fact states at any given time. Laws are changed for moral as well as for practical and political reasons, and recent years have already seen numerous changes in legal matters pertaining to informed consent (Trichter and Lewis, 1981; Simpson, 1981; Annas, 1980 and 1981). The remainder of this chapter will draw on some leading legal doctrines in discussing consent granted by one person on behalf of another, but the focus will be on the ethical justification of one or another of these positions.

The practice of one individual consenting for another has traditionally been called "proxy consent," a term often used to describe the typical situation that occurs when parents grant consent for biomedical procedures on their minor children. But that terminology is misleading, since the model of voting by proxy is inaccurate here: the person on whose behalf permission is granted (or refused) has not turned over his or her "proxy" to the "consenter." As one author notes, " 'proxy' inaccurately suggests express agency or deputization, and 'consent' connotes self-choice in medical matters" (Capron, 1982, p. 119, footnote 8). Another term sometimes used to denote this practice is "substitute judgment," but that term is also confusing because it refers strictly to one particular doctrine in which an individual is empowered to consent on behalf of another. The most neutral phrase is "third-party permission," but for the sake of brevity and continuity, the term "consent" will continue to be used in this and the following chapter.

Three leading positions drawn from other areas of the law, each having an analogue in political theories of representative government, are held to be potentially applicable to this situation (Capron, 1982; Dworkin, 1982). In answer to the question, "On what basis should one who is granted authority to decide on behalf of another make such decisions?" the following may be viewed as proper roles of the representative decision maker:

1. Substitute judgment ("the person giving permission is able to express the choices the incompetent would have made because of individualized, subjective knowledge of the incompetent" (Capron, 1982, pp. 120–121)).

2. *Best interest* ("the person giving permission will make an objectively reasonable choice that comes close to being what the incompetent, as a reasonable person, would want or that will at the least serve the incompetent's interests" (Capron, 1982, p. 121)).

3. Identity of interests ("the interests of the third party and those of the incompetent are so close that in choosing his or her own interest the third party will protect the interests of the incompetent" (Capron, 1982, p. 121)).

All three doctrines require further elucidation before their applicability in different biomedical situations can be properly assessed. For example, the notion of "best interest," although widely used and typically invoked by medical practitioners in support of their recommendations, is subject to varying interpretations and can lead to abuse of patients' rights of self-determination. When "interest" is construed in a narrowly medical sense, it is used as a rationale for instituting standard procedures and accepted therapeutic practices, the imposition of which can be supported by appeal to biomedical data about outcomes, survival rates, remissions, and the like. But when interpreted more broadly, "best interest" includes other elements that form part of a patient's value scheme, such as low tolerance of side effects, the patient's perceived quality of life, a preference for maintaining bodily integrity over an increased chance for longer survival, and additional considerations in which patients may view their own "best interest" differently from the way physicians construe it.

Are there any general conclusions to be drawn about the three senses of "representation" embedded in these legal doctrines and their applicability to third-party permission for cognitively impaired patients or research subjects? The tentative conclusions offered below take into account a range of different facts about patients, their representatives, and the biomedical procedures under consideration.

1. The substitute judgment doctrine is clearly inapplicable in the case of biomedical decisions concerning never-competent individuals: the mentally retarded or those with a lifelong history of mental disorder. There is obviously no way, in principle, of knowing what choices such persons would have made had they been competent, since there is no preexisting set of values or previously expressed wishes on which the substitute decision maker can rely. This reasoning was used by the New York Court of Appeals in its decision in the case of John Storar, although the court did not name any of the three legal doctrines as a basis of decision. John Storar was a profoundly retarded 52-year-old man suffering from terminal bladder cancer. He was receiving frequent blood transfusions as part of his treatment, transfusions that apparently caused considerable pain and suffering and that necessitated restraining the patient. Storar's 77-year-old mother, his legally appointed guardian who had initially consented to the blood transfusions, later refused further transfusions on the grounds that her son way dying anyway and the transfusions only produced increased suffering. Hospital officials brought suit to get authorization to continue treatment, which the New York Court of Appeals granted in its ruling. Without referring explicitly to the substitute judgment doctrine, the court nonetheless implicitly rejected its applicability "on the grounds that there was no realistic way to determine what John Storar himself would want done. Asking that question was, the court said, like asking, 'If it snowed all summer, would it then be winter?' " (Annas, 1981, pp. 19–20). It is worth speculating on what would have happened had the court invoked the best interests doctrine in this case. The decision might well have been the reverse, given the fact that the patient was suffering from a terminal disease and the particular treatment in question produced considerable additional pain.

2. The substitute judgment doctrine is not appropriate in cases where a patient representative is appointed by law (as in a "guardian *ad litem*") or where a legally authorized representative is someone other than a relative or long-standing friend with subjective knowledge of the incompetent. Since this doctrine can only be applied properly when the person granting permission has information about the previously

expressed wishes or the formerly held values of the patient, anyone
appointed on an *ad hoc* basis cannot fulfill this condition.

3. Substitute judgment is probably the most appropriate doctrine to
apply to elderly patients afflicted with a form or irreversible dementia,
when consent is sought from a relative with long-standing, personal
knowledge of the patient. In such cases there is virtually a lifelong set of
experiences to inform the judgment of the one giving permission. This
doctrine is also appropriate for cases of formerly competent patients
who have expressed a wish (orally or in writing), while they were still
competent, concerning what they would like to have happen if they ever
become incompetent. This was the situation of Brother Joseph Fox, an
83-year-old member of the Society of Mary, who was being maintained
on a respirator in a permanent vegetative state following a cardiac
arrest in which he suffered substantial brain damage. Prior to his
attack, he had expressed a wish orally not to be maintained by
"extraordinary means," if he were ever in a situation similar to that of
Karen Ann Quinlan. This case was decided by the New York Court of
Appeals at the same time as the Storar case. However, in the Brother
Fox case (actually called *In the Matter of Father Eichner*), the Court
allowed the patient's respirator to be turned off, at the request of a
petitioner (Father Eichner). Nevertheless, the New York Court did not
explicitly use the substitute judgment doctrine in rendering its opinion
in the case of Brother Fox, although it would have been a perfectly
appropriate use of that doctrine, and moreover, one that would have
yielded the same conclusion the court came to. Since Brother Fox had
made his wishes known while still competent, one of his colleagues,
acting in his behalf, could have rendered a "substituted judgment"
which was what in effect took place anyway, even in the absence of the
explicit use of the doctrine. To apply the best interests doctrine would
have been somewhat less clear, however, since it is questionable
whether individuals in a permanent vegetative state can properly be
said to have any "interests" at all.

4. The identity of interests doctrine is of questionable relevance in
these situations, but may be most appropriate in cases where one
spouse grants consent for treatment on behalf of another. Since
spouses frequently see their fates as bound closely to one another's,
especially with advancing age, their interests can be presumed to be
virtually identical in many cases. The question of whether an identity of
interests obtains in the case of infants and young children and their
parents, where proxy consent is required, is intriguing, but lies beyond
the scope of this essay (see Capron, 1982; Goldstein, 1982; Macklin,
1982).

5. In cases in which the best interest doctrine seems most applicable, efforts should be made to specify the particular elements believed

to constitute the patient's best interest, in an objectively verifiable manner. This task will be most difficult when relevant factors come into conflict: prolongation of life versus quality of life; restoration of mental functioning versus pain or other undesirable side effects of treatment; life prolongation of patients predicted not to return to a fully sapient state. When normally applicable criteria conflict, when some are present and others absent, it renders judgments about best interest uncertain. Furthermore, what if physicians and family members disagree in their judgments about what is in the patient's best interest? It was noted earlier that physicians often construe "best interest" in a narrowly medical sense. Family members acting on behalf of a patient are more likely to interpret their relative's best interest in light of their knowledge of the person when still competent, and this knowledge may yield a notion of "best interest" somewhat different from the narrowly medical one. But in that case, the concept of best interest begins to blur inextricably with that of substituted judgment: if family members are the best judges of a patient's interest precisely because they, the family, are most likely to be acquainted with their relative's values and previously expressed wishes, the two doctrines blend in a way that makes it hard to distinguish between them. But these conclusions are entirely philosophical, and are not intended as speculations on what courts of law might or will decide in cases where treatment decisions must be made on behalf of incompetent persons.

References

Annas, G.J. (1981). Help from the dead:The cases of Brother Fox and John Storar. *Hastings Center Rep.* **11**, 19–20.

Annas, G.J. (1980). Quinlan, Saikewicz, and now Brother Fox. *Hastings Center Rep.* **10**, 20–21.

Appelbaum, P.S., and Gutheil, T.G. (1980a). Druge refusal:A study of psychiatric inpatients. *Am. J. Psychiat.* **137**, 340–346.

Appelbaum, P.S., and Gutheil, T.G. (1980b). Rotting with their rights on:Constitutional theory and clinical reality in drug refusal by psychiatric patients. *Bull. A.A.P.L.* **7**, 306–315.

Appelbaum, P.S., and Gutheil, T.G. (1980c). The Boston state hospital case: "Involuntary mind control," the constitution and the "right to rot." *Am. J. Psychiat.* **137**, 720–723.

Appelbaum, P.S., and Roth, L.B. (1982). Competency to consent to research:A psychiatric overview. *Arch. Gen. Psychiat.* **39**, 951–958.

Bentham, J. (1789). *An introduction to the Principles of Morals and Legislation.* London.

Capron, A.M. (1982). The authority of others to decide about biomedical interventions with incompetents. In: *Who Speaks for the Child: The Problems of Proxy Consent* (Gaylin, W., and Macklin, R., eds.), pp. 115–152. Plenum, New York.

Code of Federal Regulations (1981). 45 CFR 46 Protection of human subjects. *OPRR Reports* (revised as of January 26, 1981).

Cross, A.W., and Churchill, L.R. (1982). Ethical and cultural dimensions of informed consent. *Ann. Intern. Med.* **96**, 110–113.

Donagan, A. (1977). Informed consent in therapy and experimentation. *J. Med. Philosophy* **2**, 310–327.

Dworkin, G. (1982). Consent, representation and proxy consent. In: *Who Speaks for the Child*, pp. 191–208.

Dworkin, G. (1976). Autonomy and behavior control. *Hastings Center Rep.* **6**, 23–28.

Dworkin, G. (1972). Paternalism. *Monist* **56**, 64–84.

Freedman, B. (1981). Competence, marginal and otherwise. *Int. J. Law Psychiat.* **4**, 53–72.

Goldstein, J. (1982). Medical care for the child at risk: On state supervention of parental autonomy. In: *Who Speaks for the Child*, pp. 153–188.

Kant, I. (1785). Fundamental principles of the metaphysics of morals. Originally published in 1785.

Macklin, R. (1982). Return to the best interests of the child. In: *Who Speaks for the Child*, pp. 265–301.

Meisel, A., and Roth, L. H. (1981). What we do and do not know about informed consent. *JAMA* **246**, 2473–2477.

Meisel, A., and Kabnick, L. D. (1980). Informed consent to medical treatment: An analysis of recent legislation. *Univ. Pitt. Law Rev.* **41**, 407–564.

Mill, J. S. (1863). *Utilitarianism.* London.

Miller, B. L. (1981). Autonomy and the refusal of lifesaving treatment. *Hastings Center Rep.* **11**, 22–28.

National Commission for the Protection of Human Subjects: The Belmont Report: Ethical Principles and Guidelines for the Protection of Human Subjects of Research. Department of Health, Education, and Welfare, Government Printing Office, (OS)78-0012, Washington, D.C.

Roth, L. B., Meisel, A., and Lidz, C. W. (1977). Tests of competency to consent to treatment. *Am. J. Psychiat.* **134**, 279–284.

Simpson, R. E. (1981). Informed consent: From disclosure to patient participation in medical decision-making. *Northwestern Univ. Law Rev.* **76**, 172–297.

Trichter, J. G., and Lewis, P. W. (1981). Informed consent: The three tests and a modest proposal for the reality of the patient as an individual. *South Texas Law J.* **21**, 155–170.

Veatch, R. M. (1978). Three theories of informed consent: Philosophical foundations and policy implications. In: *National Commission for the Protection of Human Subjects: The Belmont Report*, Appendix Volume. Department of Health, Education, and Welfare, Government Printing Office, Washington, D.C.

Chapter 4

Treatment Refusals: Autonomy, Paternalism, and the "Best Interest" of the Patient

Ruth Macklin*

I

Closely related to the problems of competency and consent discussed in the preceding chapter is the question of how to handle treatment refusals by patients. It was noted there that an asymmetry exists between the need physicians perceive to assess a patient's competency to grant consent in the first place and the retrospective assessment of competency once a patient has refused treatment. Put another way, questions of competency usually do not arise when a patient willingly goes along with whatever treatment the physician recommends. It is well known that treatment refusals are often construed as evidence that the patient is incompetent to grant or refuse consent, but that view of the matter is surely mistaken. However competency is to be assessed, it cannot simply be explicated in terms of agreement with what a physician recommends. Even if a patient's refusal appears irrational, on the face of it, that is still not grounds for assuming that the person is incompetent to choose or refuse a recommended treatment. It is worth recalling that presumptions of incompetency are often made when there is no other evidence of impairment—the patient is not a psychiatric patient, or suffering from senile dementia, or afflicted with a condition that normally impairs cognitive processes.

*Department of Community Health, Albert Einstein College of Medicine, Bronx, New York 10461.

What values are in tension in such cases, and how ought a resolution be sought? Before addressing that question directly, let us look first at a hypothetical, counterfactual set of circumstances that would obviate the problem of treatment refusals. A brief look at those hypothetical circumstances will help to illuminate the moral values in tension. If any one of the following three conditions obtained, the problems posed by patients who refuse recommended medical treatment could be easily resolved: (1) the only patients who refused recommended medical treatments were those who would fail to meet an agreed-upon test of competence (see preceding chapter); (2) values in our society were universally agreed to be arranged in a hierarchy such that the value of beneficence ranked higher than that of individual liberty, or the reverse; (3) being a hospital patient or submitting to the early stages of treatment obligates a patient to accept whatever further treatments a physician might prescribe. These are all counterfactual conditions, ones that are not present in the world of medical practice as we know it today in the United States. Yet it is easy to see how the problem of treatment refusals would be obviated if these hypothetical circumstances did obtain.

Under the first condition—only incompetent patients were those refusing treatment—consent refusals could be overridden by a judicial declaration of incompetence and the appointment of a guardian *ad litem*, who could then consent on behalf of the patient. If the standard of competence were set as low as possible, and if everyone agreed to accept that standard, then any patient who did not simply evidence a choice one way or the other could be treated with no further problem. The more rigorous tests of competency would be more difficult to apply, but the result would be the same: assuming that agreement could be reached both on which standard of competency should be used, and on the judgments yielded by applying that test, then if the only patients refusing treatment were those who failed to pass the relevant test, their refusal could be overridden following a declaration of incompetence. Obviously, the chief assumption in this hypothetical circumstance is false: Whatever test of competence is selected, there will be some patients who succeed in meeting that test but who nonetheless refuse treatment. This argument, in the form of a *reductio ad absurdum*, is designed to show the mistake in equating patient refusals of treatment with incompetence.

Consider the second hypothetical condition: If values in our society were hierarchically ordered such that the value of beneficence always took precedence over that of individual liberty, or the other way around, then no value conflict would arise in cases of treatment refusals. Suppose it were universally agreed that individual liberty or freedom of choice should always override beneficence when the two come into conflict. That view is very close, if not identical, to the view put forward by Thomas Szasz in a number of his writings on psychiatry and the law (Szasz, 1977; 1982). Szasz categorically opposes any form of coercion in medical and psychiatric contexts. In addition to his well-known opposition to involuntary commitment, even in the face of clear evidence that a person is dangerous to self or others, Szasz is opposed to all efforts to coerce a patient into accepting recommended medical treatment, including life-saving therapy. The extreme view Szasz holds on this issue rests on two separate bases: the first a value position, the second a theoretical stance. Szasz makes it quite clear in his written and oral pronouncements that one value stands above all others in his value perspective: individual liberty. It seems that he is prepared to accord more value to liberty than to life itself. The theoretical basis for Szasz's positon is his refusal to acknowledge a legitimate category of mental illness (Szasz, 1961). One consequence of that refusal is a failure to recognize any form of diminished responsibility for actions. It follows, then, that treatment refusals by patients must always be respected, regardless of the consequences for the patient, and regardless of judgments made by psychiatrists or others that a patient lacks the cognitive capacity to grant or to refuse treatment. The position attributed here to Thomas Szasz, and held no doubt by others as well, is internally consistent. But it requires our acceptance of two dubious premises: that freedom to choose stands above all other values, whatever the consequences; and that there is no such thing as diminished responsibility for decisions and actions.

There appears to be only one situation in which Szasz would sanction intervention where a patient has not consented to treatment: "In answer to a question about the right to intervene against a patient's will, Szasz has stated, 'It is quite obvious, and I make this abundantly clear, that I have no objection to medical intervention vis-à-vis persons who are not protesting . . . [for example,] somebody who is lying in bed catatonic and the mother wants to get him to the hospital and the ambulance shows up and he just lies there' " (Roth, Meisel, and Lidz, 1977, p. 280). This example, however, is not one in which a patient's express refusal is overridden; it is, rather, a case in which the patient expresses nothing at all.

The opposite extreme of the view Szasz holds is one in which the value of beneficence overrides all others. Accepting this value hierarchy could justify all manner of paternalistic behavior on the part of physicians: coercing patients into undergoing their proposed treatments; withholding information or even outrightly deceiving patients in order to get tham to consent to treatment; offering undue inducements; or setting the standard of competency so high that all but a few leading intellectuals would fail the test and thus be considered incompetent to grant or refuse consent for treatment. Some antipaternalistic spokes-men argue that the practice of medicine until quite recently in the United States, and even today in many places in the world, including Western European countries, embodies this value hierarchy. Whether that assessment of the paternalistic practice of medicine is true or not concerns us less than the fact that paternalistic justifications continue to be widely used in those branches of medicine in which patients evidence some impairment of cognitive function, such as psychiatry, geriatrics, and neurology.

Neither of the extreme positions on the value of liberty or bene-ficence is a reasonable one to hold. Not only is there no generally accepted value hierarchy in our society but also our cultural heritage has placed a premium on ethical and political pluralism. Furthermore, patients and their physicians may well come into conflict regarding these values central to the doctor–patient relationship. It is probably quite rare that physicians, patients, or both inquire into the other's value schemes and priority of values before entering a professional-client relationship.

The third hypothetical condition offers one way of understanding the notion of "implied consent." That hypothetical condition states that being a hospital patient or submitting to the early stages of treatment obligates a patient to accept whatever treatments a physician might prescribe. This condition should be understood to mean that physi-cians prescribe what they sincerely believe to be in their patients' best interest. If this (counterfactual) condition were in fact the case, no problems would arise concerning treatment refusals because the refusals would not count, so to speak. The patient would already have consented to everything proposed in the treatment plan. Obviously, neither the law, nor most patients, nor most physicians adopt this view of "implied consent," so the final condition that could conceivably render treatment refusals unproblematic is surely not met.

II

Since none of the counterfactual conditions discussed above obtains in the practice of medicine that most of us are familiar with, the problem of how to conceptualize treatment refusals by patients remains. In order to address the problem fully, we must take into account not only the value concepts of freedom, paternalism, and beneficence, but also the notions of autonomy, "best interest," and "standard of care." To resolve particular dilemmas regarding treatment refusals, it is necessary also to incorporate factual considerations that vary from case to case. The issue is oversimplified if construed simply as a clash between two leading values—the individual liberty or freedom of patients and the benevolent paternalism of physicians. Perhaps the most crucial concept in this area is that of autonomy—a concept more often used for rhetorical force than subjected to careful analysis. The preceding chapter contained an introduction to this topic, which must now receive careful examination. The second key concept that demands deeper analysis is "best interest," as that notion is invoked in paternalistic justifications for overriding patients' refusals. Let us look first at "best interest" before turning to autonomy.

In the previous chapter, it was noted that the concept of best interest can be construed in a narrowly medical sense, referring to outcomes stated in terms of survival months or years, chances of remission and decreased morbidity, and also in a broader sense that takes into account other values the patient may adhere to. Another way of looking at the concept of best interest gives rise to the following three questions. (1) Is there an *objective* basis on which the concept can be applied? Or (2) is it an irreducibly *subjective* notion: that which a patient believes to be the best course of action, based on his or her values and life plan? Or (3) does it represent what a physician believes to be best for a patient, based on the physician's own professional experience or on data about outcomes reported in the literature?

If "best interest" were a wholly objective notion, it should be possible for physicians and their patients to agree on a treatment plan once the diagnosis and prognosis are explained to the patient, along with relevant statistics about probable outcomes of alternative treatments (including no treatment at all). However, "best interest" is not a purely descriptive concept, although a knowledge of relevant facts is surely necessary for making judgments about what is in a person's best interest. It is a value-laden concept, which is precisely why a physician's and patient's assessment may diverge; even if they agree on all the facts of the matter, the patient and physician may still *value* the different outcomes differently. A patient may see death as a lesser evil than continued painful or debilitating therapy that diminishes quality of life. A physician may view a high risk–high gain procedure as worth pursuing, while the patient may prefer instead to suffer diminished functioning as a result of foregoing a risky recommended treatment. There is an irreducible subjectivity to judgments about "best interest," a subjectivity of value priorities that serves to explain why in many instances there is no objectively right or wrong answer to a question about what is in a person's best interest.

It would be useful to try to distinguish a class of cases in which there is arguably an objective notion of "best interest" from those in which a person's best interest is simply what he or she subjectively determines it to be. For the latter category, I propose the following cases: (a) choosing a treatment for breast cancer, e.g., lumpectomy instead of radical mastectomy, where a woman may prefer a shorter life with less mutilation to a chance of longer survival with surgery that is unacceptably disfiguring to her; (b) a patient deciding to terminate renal dialysis, after having undergone the treatment for some time (many months or years), with unremitting discomfort, nausea, fatigue, lethargy, depression, and impotence; and (c) choice of continued chemotherapy versus cessation by a patient whose quality of life is exceedingly poor and who stands to have life prolonged by, e.g., 2 years with continued chemotherapy as against only a few months without. These three cases do not exhaust the category of subjectively determined best interest, but they are paradigm cases in which the patient's determination of what is in his or her best interest is the correct interpretation of the "best interest" doctrine.

In contrast, I propose the following case as a paradigm of objectively determined best interest, in this instance, one in which the patient is incapable of making an assessment of what is in his best interest.

> Mr. D is a thirty-year-old single author who suffered an episode of mania two years ago. Since then he has been taking lithium and apparently doing well. He has just used up the small savings that had enabled him to continue writing, when he sells the novel he has been working on for the past several years. After the sales he abruptly stops taking lithium and within two months develops signs of mild to moderate hyperactivity, pressured speech, anorexia, and insomnia. Soon thereafter, a good friend of his becomes aware that Mr. D has accumulated in his apartment, in cash, the proceeds from the sale of his book as well as the small amount of money remaining in his savings account. These funds constitute Mr. D's sole means of support for the foreseeable future: the friend is astounded to learn that Mr. D, normally a careful and frugal man, has begun burning the money. Mr. D gives his friend no coherent explanation for doing so but tells him a rambling and jumbled story which involves not trusting the bank and also fearing that his home may be robbed. He wishes to destroy the money rather than let it fall into someone's hands. Mr. D's friend is alarmed by Mr. D's behavior; with some effort he persuades Mr. D to accompany him to see Mr. D's psychiatrist, Dr. L. After Dr. L interviews Mr. D and becomes familiar with the situation, he informs Mr. D that he ought to enter the hospital or at least begin taking his lithium again; meanwhile, he should let someone else safeguard his money. (Culver and Gert, 1982, pp. 171–172)

Cases of manic behavior, such as the one just described, are clear instances in which persons are unable to assess their own best interest, and it is relatively easy to provide objective criteria by way of justifying that judgment. In contrast to cases a–c above, the manic individual does not simply have a different value system from that of his friend and his psychiatrist. Rather, the values to which Mr. D adheres are not even those that he himself possessed when not in a manic state.

The objection might be raised that the case of Mr. D describes a psychiatric patient, and psychiatric patients, virtually by definition, cannot be said to know their own best interests. That objection, however, goes too far too quickly. I believe it would be a mistake to rule out by fiat the possibility that psychiatric patients may in a number of respects be the best judge of what is in their own best interest. A typical case worth considering is the trade-off between psychosis and the side effects of phenothiazines. One does not have to be a follower of the R.D. Laing school of antipsychiatry to acknowledge that a person may reasonably prefer succumbing to psychotic episodes to suffering from tardive dyskinesia, or to being unable to read because of chronic, seriously blurred vision resulting from medication.

If this attempt to divide into two categories cases in which a patient's best interest is at stake is sound, with one category described in terms of the patient's subjectively determined best interest and the second in terms of an objective concept of best interest, then an important consequence emerges for decision making about treatment. When there is no clear right or wrong answer to an ethical dilemma, the question that assumes paramount importance is the "who decides" question. It is on this very issue that the long-standing practice of paternalism in medicine has begun to give way to the exercise of patients' rights—the right to control one's own body, the "right to be let alone," the right of self-determination, and the right to refuse treatment.

In the absence of an objective concept of best interest, the situation is reduced to a procedural matter: who should decide, or where should authority to make final decisions reside? Here is where paternalists and antipaternalists divide sharply, able to settle their differences only by addressing competing rights and conflicting authority of physician and patient. One rational strategy remains, however, in an effort to resolve the impasse. That strategy requires explicating fully the concept of autonomy and reintroducing one of its richer meanings into the heart of the debate over treatment refusals.

III

To begin the discussion, let us return for a moment to the case of Mr. D described above. What is at stake in that case and others like it, when patients are coerced into accepting treatment, is not a *violation* of the person's autonomy, but rather, an attempt to *restore* it. This highlights the need to distinguish sharply the concept of autonomy from that of liberty, free action, or self-determination. Failure to make that distinction results in the oversimplified view that liberty and autonomy mean roughly the same thing, and to override a person's freedom to choose is thereby to rob him of his autonomy. It would also follow from failure to make the distinction that if a person cannot clearly be shown to be incompetent, by any of the usual tests of competency, there is no course of action left but to respect that person's liberty and allow him to refuse treatment. That that view is mistaken can be shown by appealing to one of the richer senses of autonomy, that is, by abandoning the simple equation between autonomy and free action. If liberty and autonomy are not precisely the same, then it is entirely possible to limit or override a person's liberty without at the same time violating his autonomy.

Furthermore, although in many cases, as noted earlier, a person may well be the only proper judge of what is in his or her own best interest, it is in precisely those cases in which he or she is *not* the best judge that he or she lacks genuine autonomy. This picture is complicated by the fact that to lack genuine autonomy in one of these richer senses is not necessarily to be incompetent to grant or refuse consent to treatment, according to one or more of the tests of competency described in the preceding chapter. Let us turn next to the analysis supplied by one writer who has identified at least four senses of "autonomy" as that concept is used in medical ethics.

Bruce Miller (1981) claims that four different senses of "autonomy" can be distinguished: (1) autonomy as free action; (2) autonomy as authenticity; (3) autonomy as effective deliberation; and (4) autonomy as moral reflection. The first of these is the sense we have been referring to as equivalent to liberty or self-determination. According to Miller, "autonomy as free action means an action that is voluntary and intentional. An action is voluntary if it is not the result of coercion, duress, or undue influence. An action is intentional if it is the conscious object of the actor" (Miller, 1981, p. 24).

Miller's second concept of autonomy captures half of the sense advocated by Gerald Dworkin (1976) in his formula: autonomy = authenticity + independence. Miller explicates this second sense as follows:

> Autonomy as authenticity means that an action is consistent with the person's attitudes, values, dispositions, and life plans. Roughly, the person is acting in character. . . . For an action to be labeled 'inauthentic' it has to be unusual or unexpected, relatively important in itself or its consequences, and have no apparent or proffered explanation. (Miller, 1981, p. 24)

At least two features of this sense of autonomy are worth noting for our purposes here. First, in order to judge whether or not a person is acting autonomously, it is necessary to know more about that person than the facts surrounding the action or decision in question. In order to judge whether someone is acting in character, it is necessary to have some knowledge of what that person's character traits are. In order to know whether or not an action or decision is consistent with a person's attitudes, values, and life plans, we need to have some acquaintance with the latter. That knowledge is precisely what is often lacking in the medical setting, especially where patients are under the care of house staff or a specialist who has been referred by a primary care physician. Since many patients in today's practice of medicine do not have a family physician who has known them all their lives, or even much of their adult lives, a necessary element for judging the authenticity of patient choices is often lacking.

The second feature to point out about autonomy as authenticity is that since an action or decision by a patient is not to be judged in isolation, all manner of seemingly irrational decisions may turn out to be genuinely autonomous if they are consistent with the person's character, values, or life plans. This suggests a possible shortcoming in the use of this sense of autonomy in regard to treatment refusals. In this sense of the term, a person with a lifelong history of mental illness may nonetheless turn out to have autonomy, so long as attitudes and values remain consistent, or "in character." The problem then becomes how to connect the concept of autonomy with that of competence or incompetence. Can someone judged incompetent according to one of the standards nevertheless be said to possess autonomy? There is no simple answer to that question, in part because it depends on just which of the tests of competency is chosen. As will become clear shortly, there is a closer link between competency as determined by at least two of the standards proposed by Roth, Meisel, and Lidz (1977) and autonomy in one of the remaining two senses Miller (1981) analyzes than there is between this sense (autonomy as authenticity) and any of the four standards.

The third sense of autonomy is "autonomy as effective deliberation." This means "action taken where a person believed that he or she was in a situation calling for a decision, was aware of the alternatives and the consequences of the alternatives, evaluated both, and chose an action based on that evaluation" (Miller, 1981, p. 24). This sense of autonomy bears a close resemblance to two tests of competency: factual understanding, and the next most stringent: rational manipulation of the information. Miller (1981) argues that effective deliberation is distinct from authenticity and free action, by virtue of the following considerations: "A person's action can be voluntary and intentional and not result from effective deliberation, as when one acts impulsively. Further, a person who has a rigid pattern of life acts authentically when he or she does the things we have all come to expect, but without effective deliberation" (Miller, 1981, p. 24). Miller contends that the doctrine of informed consent protects the right to autonomy in this sense of the term, since properly informed consent requires that the patient be apprised of the risks and benefits of the proposed treatment, as well as any alternative treatments that might be chosen instead of the one the physician recommends.

The final sense of "autonomy" Miller identifies is termed "autonomy as moral reflection." Just as the second sense captures one of the elements present in Dworkin's formula, "autonomy = authenticity + independence," this fourth sense includes the other half of Dworkin's formula. According to Miller,

> Autonomy as moral reflection means acceptance of the moral values one acts on. . . . One has reflected on these values and now accepts them as one's own. This sense of autonomy is deepest and most demanding when it is conceived as reflection on one's complete set of values, attitudes, and life plans. It requires rigorous self-analysis, awareness of alternative sets of values, commitment to a method for assessing them, and an ability to put them in place. (Miller, 1981, p. 25)

Given this characterization, it is questionable whether most people even possess autonomy some of the time. While the concept makes perfectly good sense analytically, it is probably more of an ideal notion of autonomy than one typically achieved in reality. A further problem with this sense is epistemological: It may be very difficult, if not impossible, for anyone to assess whether another person possesses autonomy in this sense of the term. It is difficult enough to be objective about oneself when one reflects on one's own values, attitudes, and life plans. But to determine whether another person has succeeded in doing that poses an almost insuperable epistemic barrier. Miller is careful to distinguish this last sense of autonomy from that of effective deliberation: "one can do the latter without questioning the values on which one bases the choice in a deliberation" (Miller, 1981, p. 25)

Miller uses his analysis to discuss a number of different cases of treatment refusals by patients. He concludes by noting that whether to respect a refusal of treatment requires a determination of what sense of autonomy is satisfied by a patient's refusal, and also that there need not be a sharp conflict between autonomy and medical judgment. This is because when patients refuse treatment, they may not always be autonomous in the relevant sense. This conclusion is attractive since it effects a rapprochement between two apparently opposing values:

physician paternalism and patient autonomy. That rapprochement is possible, however, only if one of the richer senses of autonomy Miller analyzes is put in place of the simplistic concept of autonomy as free action.

A somewhat similar conclusion is reached by Eric Cassell (1978), who invokes Gerald Dworkin's proposed definition of autonomy. Cassell argues that illness can and often does impair an individual's autonomy. He claims that "the sick are different than the well to a degree dependent on the person, the disease, and the circumstances in which they are sick and/or are treated . . . after all, body-image helps make up our authentic self. . . . It is clear that illness can impair authenticity" (Cassell, 1978, p. 40). Cassell also holds that illness has an effect on an individual's independence, and taking the notions of authenticity and independence together, it becomes evident that illness interferes with autonomy.

Even more than Miller, Cassell (1978) tries to show how the physician can act to respect a patient's autonomy even while failing to respect that patient's refusal of treatment. Cassell points out that the physician, in his relationship with the patient, can help restore that patient's authenticity and also independence. Arguing a somewhat unique position, he contends that "the function of medicine is to preserve autonomy and that preservation of life is neither primary nor secondary but rather subservient to the primary goal" (Cassell, 1978, p. 43). Acting so as to preserve a patient's autonomy *may* require a physician to override a patient's treatment refusal, for example, in cases where a physician may not have time or opportunity to discover whether the patient's desire to avoid treatment is authentic or not. Although Cassell's stance will no doubt be viewed as unacceptably paternalistic by extreme defenders of patient autonomy, it is very likely that those defenders continue to construe the concept of autonomy in its most simplistic interpretation: autonomy as free action, or self-determination. Furthermore, as a practitioner with an abundance of clinical experience, Cassell recognizes the omnipresence of doubt and uncertainty in the medical setting. This leads him to conclude: "It appears reasonable to me that *where doubt exists* doctors should always err on the side of preserving life. While there may not always be hope where there is life, there are usually more options" (Cassell, 1978, p. 44).

It is an unfortunate feature of modern medical practice that physicians often do not know their patients sufficiently well even to begin to ascertain whether genuine autonomy is present or not. For any but the minimalist sense of the term, an assessment of autonomy requires that a physician know something about the patient's thought processes, history, values, and life plans—if not directly, through long-standing acquaintance with the patient, at least through recent conversations with the patient or even the patient's family. These features of the physician–patient relationship are, typically, totally absent in the municipal hospital setting, where patients are cared for by young house staff with little experience and little or no knowledge of the many patients whom they care for in a largely impersonal manner. There is thus this feature of the two-class system of medicine in addition to those often remarked upon: The absence of a relationship between physician and patient makes it impossible, or at least very difficult, to determine the degree of autonomy a patient has, and whether the patient's disease has compromised his or her genuine autonomy. If we accept Cassell's conclusions, we are forced to accept greater paternalism by physicians than might otherwise be indicated, precisely because of the pervasive doubt that is bound to exist about the authenticity and independence of patients' refusals of treatment.

IV

As difficult as it is to make assessments of autonomy in patients who are physically ill but not mentally or cognitively impaired, the problems that arise with psychiatric patients are even thornier. Civil libertarians and other defenders of patient's rights hold a position regarding the right of mental patients to refuse treatment that has a stronger and weaker version. The stronger version accords patients the right to refuse treatment (typically, psychoactive drugs and electroconvulsive therapy) even in the absence of a determination that the patient is competent to grant or refuse treatment. A weaker version accepts the possibility or even likelihood that at least some psychiatric patients can pass tests of competency, and those who succeed should be granted the right of refusal. The opposing assumption, widely held by psychiatrists, is that a diagnosis of mental illness is itself a sufficient indication that a person is incompetent, and that psychiatric patients should not be permitted to refuse treatments that serve as the only hope of rendering them functioning members of society, or at least, no longer suicidal or delusional.

An increasing number of legal cases, the first two of which emerged in the late 1970s (Rogers v. Okin in Massachusetts and Rennie v. Klein in New Jersey), have granted mental patients the right to refuse psychotropic medication. These cases were taken to higher courts, and similar suits have since been brought in other states. These developments have spawned an ongoing debate in the legal and psychiatric literature, with most psychiatrists highly critical of the judicial decisions that upheld psychiatric patients' right to refuse psychoactive medication. One proponent of this viewpoint writes:

> ... the patient in the throes of a depression [which by clinical definition renders the individual's outlook on the world as 'hopeless'], is unable to make a rational decision about a projected treatment plan. His illness precludes him from seeing 'light at the end of the tunnel'—indeed, so black is his mood, that suicide for him at this time may be secretly held to be his only 'realistic' alternative. However, a court of law may view him as rational, albeit depressed, with cognitive functions intact, and capable, therefore, of accepting or rejecting a treatment plan. (Shwed, 1980, p. 194)

This author holds the overall view that the illness of severely disturbed patients precludes a rational decision regarding treatment. The question persists, however, whether consent or refusals must be *rational*, or rather, in conformity with a less rigorous standard of decision making. Shwed's assessment of such patients' inability to engage in genuine, meaningful choice is not limited to depressed or suicidal patients, as in the above-quoted passage. Discussing the manic individual, Shwed argues:

> The manic individual, who spends his day 'wheeling and dealing,' impulsively making inappropriate purchases, reducing his life savings on ill-conceived business ventures to the detriment and despair of his family, may grandiosely refuse treatment and have little or no insight into the fact that he is ill. By legal definition, however, many such patients would be judged 'competent.' (Shwed, 1980, p. 195)

No example illustrates better the tension between the right of self-determination and that of benevolently motivated protection of patients than this sort of case. Unlike patients with manifest thought disorders, or those who cannot communicate at all, the manic individual will most likely pass all of the tests of competency described by Roth, Meisel, and Lidz (1977) with the possible exception of the most stringent. On the other hand, insofar as manic behavior is episodic, it is possible to identify as the *authentic* self the individual when in a nonmanic state. By this assessment, an individual undergoing a manic episode may *pass* the test of competency, but still be *lacking* genuine autonomy. To adjust the standard of competency to fit the particular disorder manifested by different patients is to manipulate those standards in order to make them come out the way we think they ought to in each specific situation—preserving the patient's right of self-determination in some cases, and acting in accordance with the principle of beneficence in others.

In a commentary on Shwed's editorial, Laurence Tancredi (1980) makes a telling observation about the different ways in which psychiatry and the law view the concept of competency. Tancredi writes:

> Competency, a legal concept, is viewed differently by the various professions involved in the care of mental patients. The physician or psychiatrist operates under the assumption that the mere existence of disease or illness requires treatment whenever possible. Hence, there is a philosophical or value bias in the direction of questioning the competency of an individual who would not want to treat his condition. . . . The law is concerned with formulating rules for assessing competency that would thoroughly determine a specific individual's incapacity to fully understand the nature and consequence of his decisions. (Tancredi, 1980, pp. 200–201)

Tancredi is one of an increasing number of writers on this topic who holds that competency must be seen and evaluated with respect to the particular act in question. But he is somewhat skeptical of the value of the tests for competency for treatment as described by Roth, Meisel and Lidz (1977). About those tests, Tancredi says "unfortunately, none of these tests is precise enough to provide helpful guidance for the physician. On close scrutiny, it appears that they are used to justify decisions based on a cost–benefit analysis or trade-off of the treatments being proposed as they balance against the individual's rights that are affected. In the mental health care system, this includes the whole gamut of treatments from confinement to the use of physical and chemical treatments that are most intrusive and invasive . . . " (Tancredi, 1980, p. 203).

Although the question of how to deal with patients' treatment refusals has no clear and simple answer, it is nonetheless evident why this is so. It is tempting to think that the basic problem comes down to a value conflict between patients' rights of self-determination and physicians' benevolent paternalism. But as we have seen in this chapter, that straightforward value conflict is made more complex when we factor in the epistemic difficulty of assessing a patient's autonomy in particular cases: for example, is refusal of treatment truly an autonomous act, that is, authentic and independent? Still further uncertainty can be traced to the concept of competency, as discussed in the preceding chapter, and to disagreement about whether to set the standard high or low. Finally, the fact that each case possesses different individual characteristics makes it difficult to offer any sound generalizations about treatment refusals. The attendant risks and probable side effects of a proposed treatment are surely relevant to assessing the rationality of a patient's refusal, and so is the particular patient's own set of values, the nature of the illness, and the family constellation. There are thus conceptual, epistemic, factual, and ethical aspects of the problem of treatment refusals, and each aspect deserves careful attention and analysis lest the solution to such problems be arrived at too glibly or dogmatically.

A final observation is worth making. The more serious and irre-vocable the consequences of a patient's refusal of treatment, the more that efforts to persuade the patient to consent are called for. Efforts at persuasion are sometimes equated with coercion, but it would be a mistake to make that identification. On the contrary, rational means of persuasion are respectful of autonomy, of persons as reflective, deliberative agents capable of assessing the consequences of their actions and appreciating those consequences. It is a feature of modern medical practice that physicians spend too little time altogether talking with patients. In the case of treatment refusals, just as it is less than ethical for a physician to threaten or frighten the patient into accepting treatment, so too is it a moral failing simply to walk away and accept the patient's refusal at face value. Efforts at rational persuasion are not only ethically permissible but also morally obligatory. It is the surest way of showing respect for a patient's autonomy to engage in a full and frank discussion of the issues. In the happiest circumstances, patients may well come around to accepting the physician's recommendations as consistent with their own values and life plans. Of course, some patients will continue to refuse. The burden then falls to the physician to try to determine, insofar as possible, the patient's competency to consent or refuse consent, and also to assess whether the refusal is autonomously made. Those determinations are certainly not easy to make, but they represent a more conscientious approach to patient care than seeking a judicial solution to the problem. While it is true that significant gains in patients' rights regarding informed consent have come about through court decisions, it is also true that too much reliance on the courts in individual clinical decisions is not an especially good way to practice medicine. The problem of treatment refusals is not likely to disappear, but solutions in individual cases might be achieved more readily by better communications between physicians and their patients.

References

Cassell, E. J. (1978). What is the function of medicine? In: *Death and Decision* (McMullin, E., ed.), pp. 35–44. Westview Press, for AAAS, Boulder, Colorado.

Culver, C. M., and Gert, B. (1982). *Philosophy in Medicine*. Oxford University Press, New York.

Dworkin, G. (1976). Autonomy and behavior control. *Hastings Center Rep.* **6**, 23–28.

Miller, B. L. (1981). Autonomy and the refusal of lifesaving treatment. *Hastings Center Rep.* **11**, 22–28.

Roth, L. B., Meisel, A., and Lidz, C. W. (1977). Tests of competency to consent to treatment. *Am. J. Psychiat.* **134**, 279–284.

Shwed, H. (1980). Social policy and the rights of the mentally ill: Time for the re-examination. *J. Health Politics Policy Law* **5**, 193–198.

Szasz, T. S. (1977). *Psychiatric Slavery: When Confinement and Coercion Masquerade as Cure.* The Free Press, New York.

Szasz, T. S. (1961). *The Myth of Mental Illness.* Hoeber-Harper, New York.

Szasz, T. S. (1982). The right to refuse treatment: A critique. In: *Who Decides?* (Bell, N. K., ed.). Humana Press, Clifton, N.J.

Tancredi, L. R. (1980). The rights of mental patients: Weighing the interests. *J. Health Politics Policy Law* **5**, 199–204.

Chapter 5

Ethical Considerations in the Care of Unconscious Patients

DAVID E. LEVY, M.D.*

Care of the unconscious patient highlights many ethical dilemmas that face modern society. Most unconscious patients either die or recover mental function within a few days, and this rapid resolution avoids appreciable ethical problems. Many, however, linger for months or years in the vegetative state (Jennett and Plum, 1975; Levy *et al.*, 1978), in which the eyes usually open but the patient remains unaware of his environment; this vegetative state, together with prolonged coma and other related conditions, has recently (President's Commission, 1983) been designated permanent unconsciousness. Still other patients regain consciousness, but with only the rudiments of psychological awareness. All these patients pose particularly troublesome ethical issues: the problem, stated simply, is whether or not to limit life-sustaining treatment for a patient who cannot participate in the decision.

The problem is growing because the number of unconscious patients is growing. A variety of causes can probably be invoked to explain this increasing prevalence of unconscious patients. The emergence of new resuscitation technologies has made it more and more possible to preserve a functioning body in the absence or near absence of higher brain function. This trend of improved systemic care (i.e., care directed primarily to bodily rather than brain function) has been particularly

*Department of Neurology, Cornell University Medical College, New York, New York 10021.

apparent in medicine's growing ability to resuscitate patients from cardiac arrest and to sustain patients suffering various forms of severe stroke and head injury. Intensive care has thus assured the survival of increasing numbers of unconscious patients, i.e., the incidence has increased. Similarly, improved nursing care has increasingly prolonged the survival of those badly brain-damaged individuals who are "pulled through" the acute emergency. Finally, an increasingly aging population daily enlarges the numbers of patients with severe dementia for whom attentive nursing care assures almost indefinite survival.

One must emphasize that this discussion does not imply that modern technology is effective in greatly prolonging the existence of patients who are brain-dead. The hearts of patients who meet medical brain-death criteria almost never beat for more than a few days once *all* brain function has ceased (President's Commission, 1981). The comatose or vegetative patient can, on the other hand, be sustained for years if given intensive care at the outset and adequate nursing care thereafter. The record thus far is one woman who survived 37 years without regaining consciousness (*New York Times*, 1978). Chronic hospitals, nursing care facilities, and even private homes are now occupied by other unconscious patients who may in time vie for this record.

The prolonged survival of the unconscious patient imposes costs on many segments of society. First, individuals now commonly state their preference that heroic measures not be expended on them if they are only to survive in a vegetative state; that is, many people attach to such an existence a cost in terms of lost human dignity. The continued maintenance of unconscious patients places tremendous emotional and often financial burdens on families that must deal with them on a prolonged daily basis. When families are no longer able to shoulder these financial costs, they are usually absorbed either by hospitals or by society at large. Not only is this distressing to professional staffs, whose entire training has been oriented toward the goal of healing, but also the requirement of hospitals to absorb such costs is being severely strained in these times of fiscal austerity. Society (government at least) is beginning to face and retreat from the disproportionate rate of increase in medical costs. All must recognize that these financial costs are simply a reflection of other true, though possibly less tangible, costs to society. Intensive care facilities are limited. Thus, expending heroic efforts on unconscious patients who have no reasonable likelihood of recovery necessarily limits the care available to patients with better prospects.

To be balanced against these costs in reaching social and ethical decisions are the possibilities, however remote, of benefit first, to the patient if consciousness (and especially independence in daily life) were to be regained and, second, to society from respecting human life in any form whatsoever and thus, for some, making all possible efforts to preserve it. Although consideration of these potential benefits to society is beyond the scope of this chapter, all would probably agree with the importance of rendering dignified care to the human body, whether unconscious or even lifeless. On the other hand, the possibility of benefit to the patient is within the realm of clinical investigation designed to learn whether an individual unconscious patient has some, albeit small, chance of recovering consciousness and whether the patient would consider the quality of that recovery worthwhile.

Even if the quality of potential recovery can be gauged, it is often difficult to apply customary decision-making methods for "incompetent" patients to the unconscious patient. Instances where patients have discussed in advance their preferences for care should they become unconscious are rare, and thus the wishes of most unconscious patients are unknown and cannot explicitly be considered. Even where prior discussions have taken place, one can question the ability of healthy individuals previously unexposed to unconscious or seriously ill patients to appreciate intellectually and emotionally the problems serious disability can pose. Consequently, there is some doubt about the applicability of decisions made in advance. Physicians, families, and others actually caring for unconscious patients must usually rely therefore on their own judgment.

To base these difficult decisions on informed judgment requires knowledge of both the likelihood of recovery and the quality of that recovery. In so important an area, those forced to make critical decisions about continued care require specific information about recovery from unconsciousness, and the remainder of this chapter will explore two aspects of estimating prognosis from coma—first, a description of factors influencing recovery and, second, an effort to use these factors to estimate in an individual patient the likelihood of recovery.

Coma was generally regarded by the medical community as a relatively hopeless sign until the past several decades, but unconsciousness (or even coma) cannot be considered a homogeneous entity. It is in fact a heterogeneous state in terms of cause, manifestations, and prognostic important. In 1966, Plum and Posner (1966) proposed that careful analysis of the clinical picture of comatose patients could help in the important first step of diagnosing the cause of coma, with the additional important observation that diagnosis carried therapeutic implications. This approach has since achieved widespread application, with the original monograph growing through two subsequent editions in English (most recently, Plum and Posner, 1980) and translations into several foreign languages.

The evaluation of techniques to determine prognosis for patients in coma has developed somewhat more slowly but has been approached by a number of clinical investigators. Jennett and his collaborators concentrated their efforts on traumatic coma; a recent compilation of their results can be reviewed in a monograph on head trauma (Jennett and Teasdale, 1981) and an earlier report focused on prognosis (Jennett et al., 1979). Several investigators have published on prognosis from coma following cardiac arrest; two large series are those by Earnest et al. (1979) and Snyder et al. (1977). The largest effort grew from an international collaborative effort involving investigators at Cornell University Medical College in New York City, The Royal Victoria Infirmary in Newcastle-upon-Tyne, and San Francisco General Hospital; they considered prognosis from all types of nontraumatic coma with the exception of coma caused by drug overdose. The drug-overdose patients were excluded because an extensive experience (Plum and Posner, 1980) had already established that most patients who arrive in hospital comatose because of drug overdose recover completely if given good medical care. The results of this investigation of nontraumatic coma were published in 1977 (Bates et al., 1977) and 1981 (Levy et al., 1981). These papers describe the importance of individual clinical signs in estimating prognosis and further describe one effort at combining signs into a prognostic method applicable to all patients with nontraumatic coma. Some of these results will be recapitulated now, and then further extensions of this work will be discussed in greater detail.

As noted, the Cornell study concentrated on patients whose coma was caused neither by trauma nor by drugs. Only patients hospitalized on the adult service were included. In order to exclude premorbid patients or those who recovered rapidly, the protocol required that their coma last at least 6 hr. The study employed a strict operational definition of coma that emphasized the patient's lack of wakefulness (i.e., that the eyes were closed and did not open in response to stimuli) and his lack of awareness of the self or environment (i.e., that he was unable to speak or respond "meaningfully" to painful stimuli). Most of the patients were in coma as a result of acute cardiac or respiratory arrest (210 patients) or stroke (143), with additional causes being intracranial bleeding (38), liver failure (51), or other infectious or metabolic causes (58).

Nearly three-quarters of the patient population were examined within 12 hr of the onset of coma; daily neurological examinations were performed for 1 week and then twice weekly for the first month. The best clinical responses in previously specified time periods (admission, up to 1 day, 1–3 days, 3–7 days, 7–14 days, and 14–28 days) were encoded. This clinical picture was then compared to the best level of daily function attained by the patients during various follow-up periods.

Most of the patients with nontraumatic, non-drug-poisoning coma did badly (Figure 5.1). One-quarter died within the first day, and two-thirds within the first week. Only 12% of patients survived to the end of the first year. Survivors lived in a variety of clinical states. The majority by 1 year had regained independence in daily function (good recovery implying return of the preexisting level of function and moderate disability implying some disability but no interference with daily function), an intermediate, small group was severely disabled (conscious but dependent on others for daily life), and an even smaller group of patients remained vegetative. No patient remained in eyes-closed coma beyond 1 month. In addition to the 10% of total patients who survived 1 year in an independent state, 6% of the original 500 at some point during the year recovered independent function but subsequently died. Because so many of the deaths were attributable to underlying medical illness, the best functional state achieved within a 1-year period was used in subsequent analyses.

The first goal of the study was the identification of factors that were important prognostically; the development of predictive schemes then followed. Several clinical signs were importantly associated with outcome (Table 5.1), many of them even when the patient was first examined. For example, only 2% of patients without pupillary light reactions ever regained independent function, whereas 21% of those with some light reflex attained such recovery. Similarly, 4% of patients without eye closing in response to corneal stimulation regained independence in daily activities, but 22% of patients with preserved corneal reflexes showed improvement to that level. In later time periods, the discrimination afforded by the clinical examination was even finer. By the end of one day, patients lacking corneal reflexes uniformly failed to regain independent function. By the end of 3 days, the lack of pupillary light reflexes, corneal reflexes, or motor responses to pain were each associated with failure to regain independence in daily activities. By contrast, independent function was attained by approximately three-quarters of patients who by the end of 3 days spoke and were oriented to person, place, and date; or who exhibited orienting spontaneous eye movements; or who had normal eye movements in response to head turning or cold water stimulation; or who obeyed commands.

Age did not correlate significantly with recovery for patients in nontraumatic coma although it had been an important prognostic factor in traumatic coma (Jennett et al., 1979), where younger patients did relatively better than the aged. The cause of coma did carry some

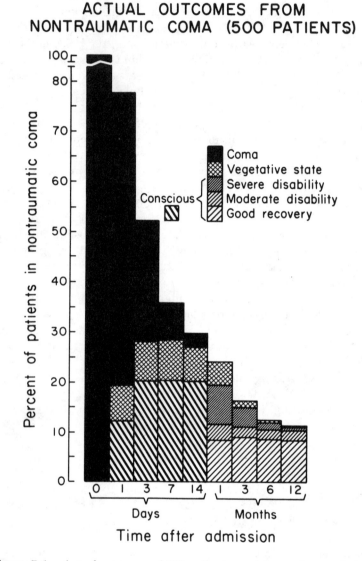

Figure 5.1. Actual outcomes of 500 patients surviving various periods after admission to the study in nontraumatic coma. Nearly half the deaths occurred within the first 3 days. Reprinted from Levy *et al.* (1981) by permission of the American College of Physicians.

Table 5.1. Moderate Disability or Good Recovery Within 1 Year Related to Early Neurologic Status[a,b]

Neurologic response	Days after admission							
	0		1		3		7	
	%	n	%	n	%	n	%	n
Verbal								
Oriented	—		67	(3)	74	(23)	83	(42)
Confused	—		64	(11)	79	(34)	82	(33)
Inappropriate	—		70	(10)	55	(11)	55	(11)
Incomprehensible	39	(57)	39	(56)	41	(27)	13	(15)
None	12	(270)	13	(176)	10	(86)	8	(39)
Eye opening								
Spontaneous	—		48	(46)	57	(107)	57	(127)
To noise	—		62	(26)	36	(22)	33	(9)
To pain	27	(101)	26	(85)	9	(32)	20	(15)
None	14	(397)	9	(227)	8	(98)	0	(26)
Pupillary light reflex								
Present	21	(362)	26	(304)	34	(232)	47	(167)
Absent	2	(134)	1	(79)	9	(26)	0	(9)
Corneal reflexes								
Present	22	(334)	29	(267)	37	(205)	48	(152)
Absent	4	(144)	0	(90)	0	(27)	0	(4)
Spontaneous eye movements								
Orienting	0	(1)	58	(24)	77	(69)	77	(84)
Roving conjugate	32	(123)	36	(132)	26	(84)	18	(55)
Roving dysconjugate	17	(35)	23	(26)	21	(14)	33	(9)
Other movement	15	(88)	7	(56)	0	(24)	8	(12)
None	9	(253)	6	(147)	2	(66)	0	(16)
Oculocephalic responses								
Normal	0	(1)	59	(32)	73	(64)	84	(79)
Full	25	(250)	23	(214)	21	(130)	14	(76)
Minimal	11	(82)	12	(49)	14	(28)	15	(13)
None	5	(166)	7	(90)	3	(36)	0	(6)
Oculovestibular responses								
Normal	20	(5)	64	(33)	72	(50)	72	(53)
Full or tonic atypical	24	(277)	20	(202)	17	(111)	17	(54)
Minimal	8	(61)	5	(38)	16	(19)	14	(7)
None	3	(122)	4	(70)	4	(28)	0	(6)

(Continued on next page)

Table 5.1 *(continued)*

Neurologic response	0 %	0 n	1 %	1 n	3 %	3 n	7 %	7 n
Motor responses (best limb)								
Obeying	—		55	(20)	73	(55)	76	(74)
Localizing	—		50	(32)	58	(38)	69	(26)
Withdrawal	27	(151)	35	(112)	24	(54)	14	(29)
Flexor	21	(73)	10	(63)	3	(32)	5	(21)
Extensor	12	(114)	7	(69)	11	(38)	0	(15)
None	7	(162)	4	(89)	0	(42)	0	(13)
Deep tendon reflexes								
Normal or decreased	19	(240)	27	(185)	35	(133)	52	(93)
Increased	19	(129)	24	(102)	34	(76)	42	(57)
None	7	(116)	6	(71)	12	(26)	25	(8)
Skeletal muscle tone								
Normal	30	(101)	45	(92)	58	(89)	71	(84)
Abnormal	16	(165)	19	(128)	21	(78)	22	(50)
None	10	(210)	6	(128)	5	(55)	5	(19)

Days after admission columns: 0, 1, 3, 7.

[a] Reprinted from Levy et al. (1981) by permission of the American College of Physicians.

[b] At each time period, the number in parentheses is the number of patients showing the specific neurologic response. The percentage indicates how many of these patients were to achieve a best 1-year functional state of moderate disability or good recovery.

prognostic weight, in that patients in coma because of stroke or intracranial bleeding did least well (only 8% regaining independent function within the first year), whereas patients with primarily metabolic causes of coma such as liver or kidney failure did best (with 32% achieving similar recovery); patients in coma because of cardiac arrest had an intermediate but poor likelihood (12%) of satisfactory recovery.

There are, however, obvious problems with such "univariate" analyses. For example, the clinician cannot easily determine the prognosis of a patient who lacks corneal reflexes but retains other relatively favorable clinical signs. This problem led to an effort to combine these various clinical signs into an easily utilized predictive scheme. The result, largely derived from a "brute force" analysis of ten clinical signs and outcome, is shown in Figure 5.2. Examination of the figure shows that the absence of several specific clinical signs, even as early as the first examination, identified a relatively large group of patients (120), only one of whom ever regained independent function, and that a similar analysis at 1 day identified 87 patients, only 2 of whom regained independence. At the other extreme, if a patient even so much as moaned on admission, he had nearly a 50% chance of achieving independent function, and the expression of any words at all by 1 day increased that likelihood of recovery to two chances in three.

This predictive technique was designed to minimize the possibility of incorrectly stating that an individual with "no chance of recovery" would in fact ultimately recover in a satisfactory manner. Otherwise, the very fact that a grim prognosis is offered could, in practice, lead to a reduced level of care that might in turn bring about the predicted outcome. Although, this predictive scheme does reduce the chance of "bad" errors to a very low level, it was developed primarily by careful inspection of the data, and better schemes might have been overlooked. Without a well-defined strategy, the development of predictive techniques becomes a serious problem as the size and complexity of the data base increase, even when a computer is available.

A recently developed recursive partitioning program (Friedman, 1977) has now been applied to the coma data in an effort to derive automatically still better predictive techniques. To implement this program fully, however, the investigator must indicate explicitly what he means by "better." The investigator has the option of assigning a weight or relative "cost" to each of the possible misclassification errors. This is a difficult task as it involves consideration of many rather diverse factors, including true financial costs, emotional burdens, and less tangible but very important concepts such as human dignity. To date, we have requested some interested neurologists at Cornell and other New York hospitals to estimate these costs, and the resulting cost matrices have been used by the recursive partitioning program to develop simple algorithms for predicting outcome from coma (Singer *et al.*, 1982).

Table 5.2 shows two representative and rather different sets of cost matrices; Table 5.3 shows the respective decision algorithms when applied to all 500 patients at admission; and Table 5.4, the resulting classification matrices. The main difference between the two cost matrices lies in the relative weights given incorrect optimistic prognoses (upper right-hand corner) and incorrect pessimistic ones (lower left-hand corner); in this context, optimistic means the prediction of recovery in a patient who never regains consciousness, and pessimistic, the prediction of no recovery in someone who does well. One of the matrices (that from physician B) assigns a relatively high cost of 20-to-1 for pessimistic vs optimistic errors; the other (A), a lower cost of only 2-to-1. Not surprisingly, these two cost matrices yielded very different

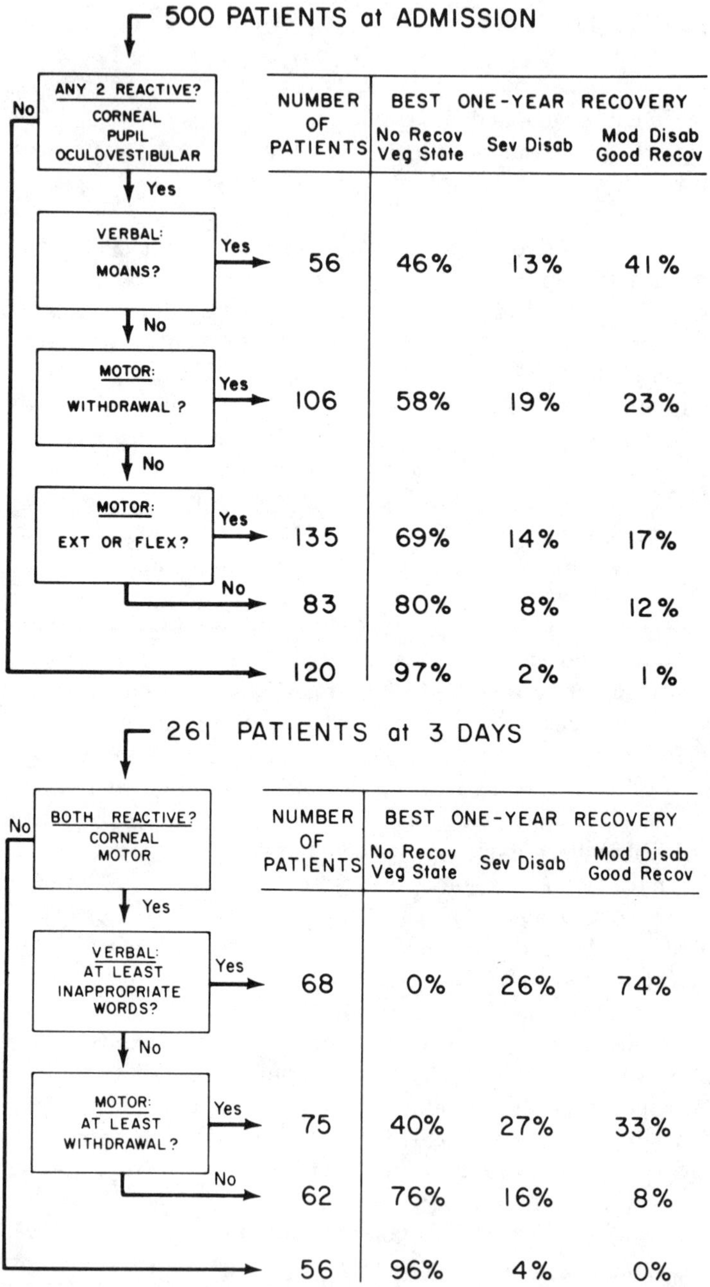

Figure 5.2. Estimating prognosis in nontraumatic coma. All patients surviving various early intervals after onset of coma are categorized on the basis of sequential criteria relating to their clinical examinations. Best levels of recovery with 1 year are given for each of the prognostic groups. No Recov = no recovery; Veg State = vegetative state; Sev Disab = severe disability; Mod

┌ 387 PATIENTS at 1 DAY

ANY 3 REACTIVE? CORNEAL PUPIL OCULOVESTIBULAR MOTOR	NUMBER OF PATIENTS	BEST ONE-YEAR RECOVERY		
		No Recov Veg State	Sev Disab	Mod Disab Good Recov
VERBAL: AT LEAST INAPPROPRIATE WORDS? — Yes	24	0%	33%	67%
MOTOR: AT LEAST WITHDRAWAL? — Yes	136	42%	21%	37%
ANY 1 PRESENT? OCULOCEPHALIC: NL OCULOVESTIBULAR:NL SPONT EYE MOVT:NL MOTOR:EXT OR FLEX — Yes	104	76%	13%	11%
No	36	84%	11%	4%
	87	98%	0%	2%

┌ 179 PATIENTS at 7 DAYS

EYE OPENING: AT LEAST TO PAIN?	NUMBER OF PATIENTS	BEST ONE-YEAR RECOVERY		
		No Recov Veg State	Sev Disab	Mod Disab Good Recov
MOTOR: AT LEAST LOCALIZING? — Yes	99	1%	24%	75%
No	54	63%	28%	10%
	26	92%	8%	0%

Disab = moderate disability; Good Recov = good recovery; Mot = motor responses; Ext = extensor; Flex = flexor; Spont Eye Movt = spontaneous eye movements; NI = normal. Nonreactive motor responses means the absence of any motor response to pain. Reprinted from Levy *et al.* (1981) by permission of the American College of Physicians.

Table 5.2. Estimated Costs of Misclassification

	Predicted Outcome		
Best 1-Year outcome[a]	No recovery Vegetative state	Severe disability	Moderate disability Good recovery
(A)			
No recovery Vegetative state	0	2	5
Severe disability	10	0	3
Moderate disability Good recovery	100	15	0

	Predicted outcome		
Best 1-Year outcome	No recovery Vegetative state	Severe disability	Moderate disability Good recovery
(B)			
No recovery Vegetative state	0	20	50
Severe disability	30	0	40
Moderate disability Good recovery	100	50	0

[a]Costs were assigned by two different neurologists (A and B) in arbitrary units on a scale of 0 to 100.

decision algorithms, and their utilization gave rise to substantially different classification matrices. The costs assigned in the first matrix resulted in a decision algorithm with many errors of prognosis—316 in all—but only one "serious" pessimistic error in the lower left-hand corner. The second cost matrix yielded much greater overall accuracy, with only 194 errors, but this improvement was gained at the expense of 17 pessimistic errors, rather than the single error of the other matrix. An examination of the actual decision algorithms in Table 5.3 shows additional influences of the cost matrix. For example, although the corneal reflex was used in both algorithms, its location in the decision process was different, and the pupillary light reflex appeared in only one algorithm, whereas verbal responses occurred in the other. Obviously exercises of this kind help the practitioner estimate prognosis in individual patients, but of equal importance, they can serve to identify ways in which physician estimates of costs influence decision algorithms, classification results, and thus medical care.

Table 5.3. Decision Algorithms

Condition[a]	Subgroup							
	1	2	3	4	5	6	7	8
(A)								
Pupillary reactions	N	N	N	Y	Y	Y	Y	Y
OC at least tonic	N	N	Y	—	—	—	—	—
OV present	N	Y	—	—	—	—	—	—
OV at least tonic	—	—	—	N	Y	Y	Y	Y
Motor at least flexor	—	—	—	—	Y	N	N	N
Any spont eye movts	—	—	—	—	—	N	Y	Y
Corneals present	—	—	—	—	—	—	N	Y
Actual and predicted (underlined) outcomes								
No recovery/ vegetative state	_88_	25	12	64	91	42	_13_	26
Severe disability	1	_4_	_1_	_10_	28	_5_	1	8
Moderate disability/ good recovery	1	1	1	7	_53_	5	0	_13_
Total	90	30	14	81	172	52	14	47

Condition	Subgroup							
	1	2	3	4	5	6	7	8
(B)								
Any spontaneous eye movts	N	N	N	Y	Y	Y	Y	Y
Motor at least w/drawal	N	Y	Y	—	—	—	—	—
OV at least tonic	—	—	—	N	Y	Y	Y	Y
Corneals present	—	N	Y	—	N	Y	Y	Y
Verbal present	—	—	—	—	—	N	Y	Y
OC at least tonic	—	—	—	—	—	—	N	Y
Actual and predicted (underlined) outcomes								
No recovery/ vegetative state	_193_	_4_	17	_36_	_18_	81	_6_	13
Severe disability	9	1	_6_	3	2	_29_	0	3
Moderate disability/ good recovery	13	0	10	0	1	35	1	_19_
Total	215	5	33	39	21	145	7	35

[a]Conditions were tested in order from top to bottom for the two decision algorithms developed from the cost matrices (A and B) shown in Table 5.2. OV = oculovestibular (caloric) responses; OC = oculocephalic (dolls eye) responses; at least = equal to or better than; Y = that the condition was present; N = either that it was not or that no information was available; hyphens (—), that the condition did not enter into the description of the specific subgroup. Underlined numbers indicate the assigned prediction for each subgroup.

Table 5.4. Correct and Incorrect Outcome Classifications

	Predicted outcome		
Best 1-Year outcome[a]	No recovery Vegetative state	Severe disability	Moderate disability Good recovery
(A)			
No recovery Vegetative state	101	146	117
Severe disability	2	17	36
Moderate disability Good recovery	1	4	66

	Predicted outcome		
Best 1-Year outcome	No recovery Vegetative state	Severe disability	Moderate disability Good recovery
(B)			
No recovery Vegetative state	253	98	13
Severe disability	17	35	3
Moderate disability Good recovery	17	45	19

[a]Numbers in each matrix correspond to the predicted and best 1-year outcomes attained as a result of the application of the decision algorithms presented for physicians A and B in Table 5.3.

Despite, or possibly because of, the importance of these cost assignments, physicians have generally encountered difficulty in attributing costs to potential errors in prediction. They are unsure about the factors to be used and the weighting to be applied to each. Among the many factors cited by neurologists in this informal survey were "the risk of possibly terminating the life of a salvageable patient," "indignity to the patient of prolonged survival in a helpless state," "the emotional cost to the family of early (or delayed) limitation of care," and "the financial cost to the family of prolonged care." Precisely how these and other factors should be considered remains an important but unresolved problem.

The rational management of comatose patients thus necessarily requires consideration by many groups concerned with health care and societal ethics. These include not only clinical investigators with their data about the recovery of previously studied patients from coma but also ethicists and others in a position to evaluate the costs and benefits to the patient and to society of various functional states and management decisions. This is not an easy task. The incorporation of ethical and medical information is, however, and endeavor worthy of careful, coordinated effort because of the importance to everyone of these life-and-death decisions.

Acknowledgments

The author is indebted to Dr. Fred Plum for his stimulating guidance into the world of ethically important decisions in neurology and to Drs. Robin P. Knill-Jones, Burton Singer, Robert Lapinski, Halina Frydman, and others whose collaboration in evaluating our coma data base has already been acknowledged (Bates et al., 1977; Levy et al., 1981). Additional important support was derived from a grant by the Robert Wood Johnson Foundation (6024) supporting the development of a computerized neurology data bank, an Established Investigatorship from the American Heart Association with funds contributed in part by the New York Heart Association, and Grant NS-42328 from the National Institute of Neurological and Communicative Disorders and Stroke.

References

Bates, D., Caronna, J. J., Cartlidge, N. E. F., Knill-Jones, R. P., Levy D. E., Shaw, D. A., and Plum, F. (1977). A prospective study of nontraumatic coma: Methods and results in 310 patients. *Ann. Neurol.* **2**, 211–220.

Earnest, M. P., Breckinridge, J. C., Raynell, P. R., and Oliva, P. B. (1979). Quality of survival after out-of-hospital cardiac arrest: Predictive value of early neurologic evaluation. *Neurology* **29**, 56–60.

Friedman, J. H. (1977). A recursive partitioning decision rule for nonparametric classification. *IEEE Trans. Comp.* **C-26**, 404–408.

Jennett, B., and Plum, F. (1975). Persistent vegetative state after brain damage. A syndrome in search of a name. *Lancet* **1**, 734–737.

Jennett, B., and Teasdale, G. (1981). *Management of Head Injuries*. F. A. Davis, Philadelphia.

Jennett, B., Teasdale, G., Braakman, R., Minderhoud, J., Heiden, J., and Kurze, T. (1979). Prognosis of patients with severe head injury. *Neurosurgery* **4**, 283–289.

Levy, D. E., Bates, D., Caronna, J. J., Cartlidge, N. E. F., Knill-Jones, R. P., Lapinski, R. H., Singer, B. H., Shaw, D. A., and Plum, F. (1981). Prognosis in nontraumatic coma. *Ann. Int. Med.* **94**, 293–301.

Levy, D. E., Knill-Jones, R. P., and Plum, F. (1978). The vegetative state and its prognosis following nontraumatic coma. *Ann. N.Y. Acad. Sci.* **315**, 293–306.

New York Times (1978). November 27.

Plum, F., and Posner, J. B. (1966). *The Diagnosis of Stupor and Coma*, 1st ed. F. A. Davis, Philadelphia.

Plum, F., and Posner J. B. (1980). *The Diagnosis of Stupor and Coma*, 3rd ed. F. A. Davis, Philadelphia.

President's Commission for the Study of Ethical Problems in Medicine and Biomedical and Behavioral Research (1981). *Defining Death: A Report on the Medical, Legal and Ethical Issues in the Determination of Death.* U.S. Government Printing Office, Washington D.C.

President's Commission for the Study of Ethical Problems in Medicine and Biomedical and Behavioral Research (1983). *Deciding to Forego Life-Sustaining Treatment.* U.S. Government Printing Office, Washington D.C. In preparation.

Singer, B. H., Frydman, H., and Levy, D. E. (1982). Sensitivity of prognostic indicators to clinician designations of tolerable misclassification errors. Presented before the Society for Medical Decision Making, Boston, Massachusetts. October 26.

Snyder, B. D., Ramirez-Lassepas, M., and Lippert, D. M. (1977). Neurologic status and prognosis after cardiopulmonary arrest: I. A retrospective study. *Neurology (Minneap.)* **27**, 807–811.

Chapter 6

Legal Aspects of Ethics in the Neural and Behavioral Sciences

H. RICHARD BERESFORD, M.D., J.D.

Introduction

A major ethical issue in clinical neurology is deciding when it is proper to withhold or withdraw life-sustaining medical care from a neurologically impaired patient. The question arises most often with respect to the irreversibly comatose and children with severe mental retardation or irremediable birth defects. It may also surface in connection with various progressive or untreatable neurological diseases, some of which destroy cognitive functions (e.g., Alzheimer Disease) or which leave cognition intact but are otherwise totally incapacitating (e.g., the motor neuron diseases). Most persons subject to ethical inquiries of this sort are unable to participate in a dialogue about what is "right" for them. They lack capacity to comprehend in even the most rudimentary fashion the nature of their condition or their prospects for the future. Others must decide what level of medical care should be provided for them. As to those few who can know their prognosis and express themselves, their wishes must be heeded. A decision which excludes them would appear ethically intolerable and is legally impermissible [1,2].

Several recent judicial decisions have considered the issue of how and by whom choices to withhold or withdraw care should be made.

*Departments of Neurology, North Shore University Hospital, Manhassett, New York 11030 and Cornell University Medical College, New York, New York 10021.

These cases highlight the complexity of the ethical problem, and are essential to an understanding of how courts determine what is lawful in the absence of legislation or other explicit social consensus. As will be seen, courts have voiced sharply differing opinions about how to reach decisions to change levels of care. But they have agreed in principle that it is lawful to withhold or withdraw life support from some persons with severe, intractable neurological impairment. In so doing, they have dealt with concepts of autonomy, constitutional and other individual legal rights, euthanasia, homicide, decision making by surrogates, and the limits of judicial competence. The doctrinal and other controversies which have emerged will occupy lawyers, judges, physicians, ethicists, and others for some years to come.

Through discussion of selected recent judicial decisions, this chapter will consider the law relating to withholding or withdrawing life support. It will also explore how law and medicine as social institutions can join to reach decisions that are both compassionate and defensible in terms of current medical knowledge. While I do not wish to tarnish others with my ideas, I must acknowledge the influences of Dr. Fred Plum of Cornell University Medical College through his comprehensive efforts to quantify predictions about outcome of severe neurological impairment, and of Professor Larry Palmer of Cornell Law School for his many insights into the potential for constructive interactions between law and medicine in handling difficult ethical issues.

Legal Aspects of Withholding or Withdrawing Care

Palliative Treatment of Fatal Illnesses

The Saikewicz Decision [3]

The central issue in this Massachusett's case was the power of institutional caretakers to withhold chemotherapy from a neurologically impaired adult with acute leukemia. The patient was a severely retarded man who had spent most of his 69 yars in custodial facilities. His IQ was scored at 10 and his behavior was that of a 2- to 3-year-old child. When he developed leukemia in 1976, his attending physicians recommended that chemotherapy be withheld. The superintendent of the caretaking institution then asked a probate court to appoint a guardian to protect his interests. When contacted, his two adult sisters declined to participate in any decisions about his care. The appointed guardian also recommended that chemotherapy be withheld. At a hearing on the propriety of withholding treatment, extensive medical evidence was introduced. It disclosed that chemotherapy might at best produce a remission for several months, but would not be curative and had many distressing side effects. There was testimony describing the inability of the patient to understand the nature or purposes of treatment or to cooperate in a complex chemotherapy program. There was also testimony that most mentally competent adults with this particular form of leukemia elect chemotherapy, knowing that it will not cure and is likely to be stressful. The judge ordered that chemotherapy be withheld, and the patient died a few months later.

The state supreme court reviewed this order and concluded that the probate judge had acted correctly. It reasoned that since mentally competent adults have a lawful right to decline life-sustaining treatment, a similar right must be accorded to mentally incompetent adults. This right is exercisable under the doctrine of "substituted judgment," whereby an apointed guardian determines what a reasonable person in the patient's position would choose and then makes a choice for the patient. As to future cases of this nature, the court declared that the proper procedure is for a probate court to appoint a guardian for the patient, hold a hearing on the guardian's recommendations, and then issue an order which either requires treatment or authorizes withholding it. The court's view was that the judicial system is more likely to produce a sound result than the unregulated deliberations of families and physicians, whether or not assisted by institutional review or ethics committees.

The Storar Decision [4]

Like Mr. Saikewicz, John Storar was a severely retarded adult who had been confined to a state institution (in New York) for most of his life. He developed bladder cancer which spread and resisted various treatments. As a result of bleeding from the bladder tumor he became anemic. To correct the anemia and ease discomfort, his attending physicians administered blood transfusions. After several transfusions, his mother requested that they be stopped. When this request was rejected, she sought a court order directing that further transfusions be withheld. New York's highest court held that the institution was not required to honor her request. It reasoned that a court should "not allow an incompetent patient to bleed to death because someone, even someone as close as a parent or sibling feels that this is best for one with an incurable disease." The court emphasized that Mr. Storar was unable to express a coherent choice, and that there was no way of determining his views. It was thus unwilling to apply the doctrine of substituted judgment to implement his right to refuse life-prolonging treatment.

Resuscitation of the Hopelessly Ill

The Dinnerstein Decision [5]

In this case, the family (including a physician-son) of an elderly woman with Alzheimer Disease agreed with her attending physician that she should not be resuscitated in the event of a cardiac or respiratory emergency. She was severely demented, incapable of self-care, had previously experienced a stroke, and had high blood pressure and heart disease. Attorneys for the Massachusetts hospital where she was a patient were concerned that the rule of *Saikewicz* required authorization from a court before an order not to resuscitate (no code order) could be implemented. An intermediate level appeals court concluded, however, that *Saikewicz* did not apply where treatment has no "reasonable expectation" of providing "permanent or temporary cure of or relief of illness."

Custody of a Minor [6]

This Massachusetts case involved an infant who had been abandoned by her parents at birth and was a ward of the state Department of Social Services (DSS). She had severe cyanotic congenital heart disease, for which one attempt at surgical repair was unsuccessful. In late September 1981, at age 4½ months, she developed an infection and was hospitalized. In early October 1981, she was placed on a respirator. After concluding that no cure of her condition was possible and that she was unlikely to live beyond 1 year of age no matter what the treatment, her attending physicians recommended that a "no-code" order be written. The effect of this order would have been to bar "heroic medical efforts," such as intracardiac epinephrine, open chest cardiac massage, or electroconversion.

The DSS and the child's guardian refused to consent to a "no-code" order and her physicians asked the Juvenile Court to authorize it. After a hearing on October 8, 1981, the court authorized the hospital administrator to enter the order, but also directed that other treatments be continued. While an appeal was pending, the child's attending physicians changed their opinion about the "no-code" order and petitioned the court to revoke its earlier authorization because of a "change in the child's condition." After further hearings, which established that the child's development and intelligence were normal for her age, the court continued in effect the "no-code" order and authorized discharge from the hospital.

The state's highest court upheld this decision on the basis of *Saikewicz*. It found *Dinnerstein* inapplicable since there was "no loving family to speak for the child," and concluded that the best approach is a judicial determination in accordance with the doctrine of substituted judgment. It defined the central issue as the manner of the child's dying, and decided that the Juvenile Court properly examined the child's best interests in applying substituted judgment. It rejected the argument that the caretakers must establish beyond a reasonable doubt that the child would have refused treatment if competent. In the court's view, adopting such a high standard of proof would cause those terminally ill patients involved to suffer unnecessary pain and loss of dignity.

Life Support Systems for the Irreversibly Comatose

The Quinlan Decision [7]

For several months in 1976, the plight of Karen Quinlan received wide publicity. She is a young woman who inexplicably lost consciousness and was placed on a mechanical respirator because of depressed respirations. She entered a persistent vegetative state [8], in which she was able to sustain certain vital functions and exhibit random movements but demonstrated no detectable awareness of her environment. When attempts were made to remove the respirator, her blood oxygen levels fluctuated to a degree that her physicians concluded that she might die without it. After several months in this condition, her father asked that the respirator be discontinued. When her attending physicians refused, he sought a court order appointing him guardian with explicit power to authorize removal of the respirator.

At a hearing before a lower court judge, there was abundant medical testimony that she was unlikely to regain cognitive functions. The judge concluded, however, that the decision about the respirator was a medical one and denied the father's petition [9]. On appeal, the New Jersey supreme court ruled that the father should be given the authority he sought as a means of protecting her constitutional right of privacy, provided that her attending physicians and a hospital ethics committee agreed that she had "no reasonable possibility of regaining cognition or sapience." The court stated that under these circumstances the participants in a decision to remove the respirator would be immune from civil or criminal liability. After the court's decision was announced, the respirator was removed. She continued to breathe without it, however, and as of this writing remains alive but comatose.

The Eichner Decision [4]

In a companion case to that of John Storar, New York's highest court was asked to rule on whether a respirator should be removed from an elderly priest, Brother Fox, who was in a vegetative state following resuscitation from a cardiac arrest that occurred during elective surgery. The medical evidence was that he had no reasonable prospect of regaining cognitive functions. Fellow priests testified that he had declared, during their discussions about the Karen Quinlan case a few years previously, that he would want a respirator removed if he were in her state. He died while litigation was in process and while still on the respirator. The court nevertheless ruled on the legal issue, and held that it is proper in these circumstances to appoint a guardian to direct removal of a respirator. For future cases of this nature, it declared that before a respirator or other life support can be withdrawn, there must be "clear and convincing" proof both that a person will not regain cognition and that he has previously expressed a desire not to have life-prolonging treatment. In other words, there must be medical evidence about prognosis and other evidence which demonstrates the person's own views in the matter. The court suggested that if such evidence exists a court proceeding will be unnecessary.

Decisional Models

General

While courts have not spoken in terms of models of decision making in these cases, they have prescribed certain procedures for protecting individual legal rights or interests. How formal these procedures should be is viewed differently by various courts. Little legislative guidance exists for judges faced with decisions about withdrawing or withholding care, and this may partly account for the differing approaches. For purposes of discussion, I have chosen to call these approaches models.

The Traditional Medical Model

Testimony from physicians in the *Quinlan* case disclosed a medical tradition of adjusting levels of care to fit the nature and prognosis of patients' illnesses. This tradition included discontinuing life-support and other treatments for patients experiencing painful, incurable illnesses. It rested on the view that one aspect of the physician's role is to ease the process of dying where neither death nor profound suffering are preventable. The trial judge in the *Quinlan* case found no legal basis for compelling Karen's physicians to remove the respirator and left it to them to decide whether removing it was a medically appropriate step. The state supreme court, however, concluded that the physicians' nonaltruistic concerns with potential malpractice or criminal liability inhibited them from making a proper decision. It thus brought her father and an ethics committee into the decisional process to provide someone to speak for the patient and perhaps to provide a measure of objectivity about the ethical questions.

It would be unrealistic to assert that a model which accords physicians' unfettered authority to withdraw or withhold care will ever again be acceptable in this country. Even those who believe that physicians are generally conscientious and knowledgeable enough to exercise this power may worry that a few careless or incompetent practitioners will cause enormous harm. The traditional medical model can only endure to the extent that the physician shares decision-making responsibility with others. The physician may still dominate the process, but someone else must speak for the patient. If the physician is the preeminent decision maker, however, he or she will remain vulnerable to legal action, at the behest of either a disaffected family member or a local prosecutor. Because of this vulnerability, one would not expect a physician to be cavalier about prognosis or too quick to withdraw life support. On the contrary, as the *Quinlan* supreme court opinion suggests, the self-protective concerns of physicians might create a bias toward unreasonable or inhumane prolonging of vegetative or tortured lives. If so, those who see unacceptable risks to patients in assigning the physician the major responsibility for setting levels of care may find the alternatives even more intolerable.

Because the traditional model is frankly paternalistic, it runs against legal and philosophical notions that individual autonomy must be protected at virtually any cost. If a patient indeed has a legally enforceable right to die, the issue is not what the physician or anyone else thinks is proper but what the patient desires. Yet for the great bulk of neurologically impaired patients in whom withdrawing or withholding treatment is in question, determining what the patient desires is simply impossible. The exceptions are those with intact cognition despite profound motor or other impairments, and the rare person who has precisely indicated while competent what level of care he or she would prefer in the event of a particular illness or incapacity (e.g., Brother Fox). These exceptions aside, one is left with choosing between the approach of the New York court in *Storar*, which rejects the proposition that someone other than the patient can authorize withdrawal of care, and the approaches of the *Quinlan* or *Saikewicz* courts which would permit persons other than the patient to decide.

Substituted Judgment

The Quinlan Model

The New Jersey court would allow family, attending physician, and a hospital committee to reach a joint agreement to withdraw care from an irreversibly comatose or vegetative patient. The participants would be immune from civil or criminal liability, provided the medical evidence is that the patient has no reasonable possibility or regaining cognition or sapience. Until the *Quinlan* decision, the scope of a physician's lawful authority to withhold or withdraw care was ill-defined. Those New Jersey physicians who, before *Quinlan*, allowed hopelessly ill patients to die through omission of certain treatments had no advance assurance that they would not be targets of legal action. On the other hand, they may have sensed that the risk of liability was so small as to be practically insignificant, particularly if they had a good relationship with the patient's family. The attending physicians in the *Quinlan* case have stated that concern over liability did not influence their decisions about use of the respirator [10]. But the state supreme court gave little credence to this in formulating its procedure for tripartite decision making.

The *Quinlan* model keeps deliberations within the hospital or other health care facility and avoids the need for a court order in each individual case. While the model envisions use of a committee, the principal purpose of the committee is probably to validate the medical prognosis. The court's opinion refers to an "ethics committee," but New Jersey lawyers and hospital officials have apparently interpreted the role of the committee more narrowly. If a committee is to review the ethical aspects of a decision to withdraw care, its composition might be quite different if its only charter is to review the medical data. Thus, a full-blown ethics committee might include persons with formal training in ethics, philosophy, theology, sociology, and law if its goal is to rationalize a decision in ethical terms. It would include only neurologically oriented physicians if determining the accuracy of prognosis is its major purpose.

Assuming that a committee is to consider only prognosis, the *Quinlan* model envisions that physicians (including attendings and committee members) will determine what will happen to a patient if treatment is continued or withheld. Then the attending physicians and the patient's family will decide what ought to be done. In this paradigm, the ethical aspects are explored in the dialogue between family and physician and in the family's own deliberations. How explicitly the ethical issues are faced may vary considerably. But it does not follow that a comprehensive ethical colloquy is essential to an appropriate decision. The patient's family is more likely to know the patient's preference than anyone else, and the physicians will have or can obtain more data about the patient's medical status and prognosis than anyone else. Whether more is needed to reach an "ethical" decision is open to question [11].

The Saikewicz Model

The answer of the Massachusetts high court is that families and physicians, whether or not assisted by committees, are not equal to the task. In the view of the court, only a judicial proceeding assures a careful factual inquiry and the needed measure of objectivity. One exception, exemplified by *Dinnerstein*, is where a patient has a fatal illness for which there is no useful treatment and where family and attending physician agree on the propriety of withholding life support. The risk of relying on the *Dinnerstein* exception, however, is that the case was decided by an intermediate level appeals court in Massachusetts and is not a bindng precedent even in Massachusetts. In other words, despite the result in *Dinnerstein* and despite considerable medical acceptance of the idea that "no-code" orders are proper for some terminally ill and suffering patients, *Saikewicz* may still be invoked as authority for requiring judicial approval of all decisions to withhold or withdraw life support from neurologically impaired, legally incompetent patients.

The disadvantages of the *Saikewicz* model are manifest. It shifts a sensitive issue of professional medical judgment from a medical forum to a legal forum which has no special expertise in such matters. It ensnarls an exquisitely private negotiation in legalisms. It rests on the unsubstantiated assumption that families and physicians are likely to ignore the interests of dying patients in a nonaltruistic fashion. It assumes that a judicial proceeding will promote "objectivity" in a situation where demonstrably correct answers are unattainable. In a very practical way, it risks encouraging physicians to provide more prolonged and intensive care for dying patients than is medically reasonable pending a judicial resolution.

Despite these shortcomings, the *Saikewicz* model may be useful in some situations. Mr. Saikewicz himself is an example. He could not speak for himself; his only family members chose not to speak for him; and his institutional caretakers were uneasy about denying him palliative treatment for a fatal disease. They then sought judicial guidance. In the course of the judicial proceeding, the nature of his illness, his capacity to accept treatment, and the risks and benefits of treatment were explored in some detail. The eventual decision to withhold chemotherapy was based on a careful weighing of the medical data and an attempt to identify and protect his interests. The judicial proceeding thus served to protect a helpless man who lacked family or friends who would push for a thorough medical appraisal or who cared enough to try to identify and assert what seemed right for him.

Even if one views judicial participation as essential, it may not be necessary for a court to conduct a full-scale hearing. In *Saikewicz*, the court itself weighed the medical evidence relating to disease, treatment, and prognosis in reaching its decision. But a different approach is possible. In response to a petition from a family member, a physician, or a legal representative, a court might appoint a guardian with instructions to ascertain from the attending physicians or any other knowledgeable physicians whom the guardian wishes to consult what is the patient's prognosis. The guardian may then present a summary of medical conclusions to the court. This could be accomplished nearly as quickly as the conventional clinical consultation in which an attending physician seeks the opinion of an expert consultant on a difficult case. Armed with the medical data the guardian has secured, the court could then ascertain the views of the family members or legal representatives about what they think ought to be done. If there is general agreement that withholding or withdrawing life support is proper, the court may promptly grant the requested authorization. No further hearing would be necessary. This approach allows judicial supervision of the decision, but places the burden of resolving the medical and ethical issues on physicians and those who represent the patient. The court would take a more active role only if disagreements or major ambivalences emerged.

The Patient-Autonomy Model

In its decision in *Storar*, New York's highest court refused to apply the substituted judgment doctrine to authorize withdrawal of life support.

It was willing to permit removal of a respirator where a patient (Brother Fox) had clearly made known his preference at a time when he was mentally competent. But it was unwilling to allow life-prolonging blood transfusions to be withheld from a patient (John Storar) who was and always had been incapable of expressing a preference. The court emphasized that every person has a right at common law to refuse any form of treatment and that physicians and others must respect this choice. Thus, when the proof was "clear and convincing" that Brother Fox would have rejected life support and that there was no reasonable possibility of neurologic recovery, the court empowered his representative to exercise his right to refuse treatment. This did not amount to a substituted judgment since the representative's only role was to implement the patient's choice.

The *Storar* opinion frankly discourages use of courts to resolve issues relating to removal of life support. At the same time, it encourages the state legislature to involve itself in establishing guidelines for reaching decisions on this question. While several states have considered "right to die" legislation and a few have enacted such laws (e.g., the California Natural Death Act) [12], the New York legislature has thus far refrained from acting. Even in the less controversial area of "brain-death" legislation, New York has stood aside. Whether this conservatism is a sensible response to complex issues or is a deplorable evasion of legislative responsibility is a matter for debate. But the difficulties in drafting workable and broadly acceptable legislation in these areas are substantial [13], and it is not self-evident that legislation will ease the resolution of particular problems, including uncertainties about neurologic prognoses and ambivalent attitudes of caretakers.

By refusing to adopt a substituted judgment rationale, the *Storar* court avoided some of the analytic problems faced by the *Quinlan* and *Saikewicz* courts. In *Quinlan*, the New Jersey court tried to put itself in Karen's place and predict what choice she would have made if she only knew her predicament. It assumed that she would have asked for removal of the respirator. But, unlike Brother Fox, no evidence was presented that she had ever envisioned herself in an irreversibly comatose state. Nevertheless, the court concluded that discontinuing the respirator would accord with her wishes, undiscoverable though they were. In *Saikewicz*, the Massachusetts court named a guardian to determine what choice the patient would make, based on the guardian's appraisal of what someone in the patient's position would choose. By contrast, the *Storar* court concluded that no persons—be they parent, guardian, or judge—were capable of knowing what the patient wanted. Thus, it did not try to imagine what he would prefer nor allow his mother to speak for him.

If the concept of personal autonomy means that only the one whose life is at stake can reject life-sustaining medical care, the implications for most comatose or other legally incompetent patients are either that there is no lawful basis for removing life support or that the physician has a measure of lawful discretion to reduce levels of care. It is improbable that many physicians would be comfortable with the notion that they have some ill-defined authority to remove life support in some ill-defined circumstances. While some may believe that the decision properly belongs to the physician, the nonmedical overtones and the possibility of later legal entanglements are disincentives to unilateral action. Thus, a potentially undesirable outcome of leaving the physician to decide is an excessive use of life-support technology. The physician who perceives a major legal risk in removing life support is unlikely to discontinue a respirator from a living but comatose patient, no matter what his beliefs or values.

The Physician-Immunity Model

The problem of a physician's self-protective reluctance to remove life support might be resolved by enacting legislation which confers immunity from civil or criminal liability. A statute might, for example, grant immunity to the physician who acts in good faith and without negligence. The physician could establish good faith by showing that he or she acted in accord with the wishes of a patient's family, and could avoid allegations of negligence by securing the opinion of a highly qualified consultant that the patient had no reasonable prospect of neurologic recovery. Or, as intimated in the *Storar* opinion, legislation might establish a series of procedures for physicians to follow in arriving at a decision to remove life support. If the procedures were followed, statutory immunity would apply.

A model which both respects personal autonomy in choosing levels of care and frees physicians from a paralyzing preoccupation with legal questions seems attractive. It would satisfy proponents of a "right to die" or reject unwanted medical care, and would comfort those physicians who feel that law forces them to act inhumanely or contrary to their best medical judgment. In *Storar*, Brother Fox's right to decline a respirator was upheld, and the physician's legal responsibility for removing the respirator was determined only by reference to the medical "facts" (i.e., the prognosis for neurologic recovery). The rub is that Brother Fox was unlike most patients who become irreversibly comatose. He had anticipated his situation and had precisely stated his position on use of a respirator. The New York court simply implemented his choice and did not consider the thorny questions of how to decide what a comatose patient wants or what is in his best interests.

For these persons who have never spoken to the issue of life support should they become irreversibly comatose, either through a "living will" or otherwise, the *Storar* concept of personal autonomy has little meaning. They are unable to speak and unable to be spoken for. If there is to be a decision maker, it must be the physician. Yet the physician may question his or her legal authority to decide.

To illustrate how the notion of legal immunity for discretionary acts might influence a physician's conduct, consider a New York physician who has concluded that a patient's coma is irreversible. Under the *Storar* standards, the physician can lawfully remove life support if the evidence is "clear and convincing" that this is what the patient wants. But suppose the patient has never spoken to the issue of use of a respirator. Here the physician has no express legal authority for removing life support. He can either continue it knowing that it is futile or can discontinue it at the behest of the patient's family or representative. If New York had a statute which provided legal immunity to physicians for actions based on a reasonable exercise of medical judgment, the physician might be encouraged to choose removal of care. Of course New York does not have such a statute. The physician who is inclined to remove life support must find other legal support for his or her actions. One alternative is to seek a court order. This is the thrust of the *Saikewicz* decision. Alternatively, the physician may choose to rely on an opinion from a lawyer that the legal risks of removing care are insubstantial. For the physician who can tolerate a mearsure of uncertainty about the legal consequences of removing care with the consent of family or legal representative, there is the option of acting without any advance legal assurances. Here the physician gambles that no court will find such a decision legally impermissible.

Conclusions and Reflections

One should not let divergent approaches of courts to the problem of withholding or withdrawing care from the neurologically impaired obscure an important judicial consensus. This is that the law permits physicians—under carefully specified circumstances—either to end the life of a patient by removing life support or to withhold care that might extend life. Given this consensus and a pervasive social concern with the rights of the dying and hopelessly ill, law and medicine have an opportunity to develop mutually acceptable approaches to the problem. Polemics about "medical paternalism" [14] and "judicial imperialism" [15] are not constructive, however much they pique the interests of readers and listeners.

Both law and medicine value fact-finding, even though their practitioners may define what is a "fact" in different ways. There is presumably agreement between the two disciplines that no decision about the care of a particular patient should be made until his condition has been thoroughly investigated and his prognosis has been as carefully defined as existing knowledge allows. This type of fact-finding can only be conducted by physicians since it requires the methodologies of medicine and biomedical science. The next level of fact-finding is attempting to ascertain what a patient wants. Characterizing the level of a patient's cognitive function is the province of the physician. But once this evaluation has been accomplished, interpretations of the patient's expressions of preference may involve both physicians and others.

Once fact-finders have agreed on diagnosis, prognosis, and the patient's preferences (if determinable), the question of how to implement a change in the level of care comes to the fore. Various approaches are possible. At one extreme is a decision by a physician, acting alone or with the advice of medical consultants. At the other pole is a full-scale courtroom proceeding, with all parties (patient, family, physician, and hospital) being represented by lawyers. An intermediate approach is that envisioned by the *Quinlan* court, involving family, physician, and a hospital committee, but omitting a judicial proceeding. Advocates of leaving the decision entirely to the physician are few, and it seems unlikely at present that many physicians are willing to assume such complete responsibility. Thus, discussion can focus on whether judicial approval of a decision to withhold care is essential.

My own preference is for a process that is hospital-based and which allocates the responsibility to attending physician and family. Fact-finding can be conducted in accordance with currently accepted standards of medical practice, and need not require elaborate rule making. In large teaching hospitals, where extensive evaluations and liberal use of consultants are routine, it seems safe to assume that diagnoses and prognoses of neurologically impaired patients will be as reliable as the state of the art permits. In smaller hospitals or those lacking staff members with neurological expertise, it may be desirable to establish rules which will assure that neurologically impaired patients are fully evaluated, even if this means seeking outside consultations. Whatever rules or procedures hospitals mandate, prevailing clinical and ethical standards require that consultations be obtained in problematic cases. There is also the factor that hospitals and physicians may be subject to civil or criminal liability if they improperly withhold or withdraw medical care. Taken together, these various elements provide strong incentives to avoid mistakes in diagnosis and prognosis. According the family a central role in the decision-making process assures that someone will speak for the patient, and should reduce the concern of those who believe that physicians may not be sufficiently sensitive to the "human" factors.

A favored argument for judicial participation is the need for objectivity [14]. In terminal care decisions, the search for objectivity may encompass both the quality of medical evaluations and the rationale for withholding or withdrawing care. Courts, however, are not especially well equipped to assess the qualitative aspects of a physician's performance, unless extensive medical testimony is available (as in a malpractice case). Moreover, most decisions of this nature are nonadversarial; if physician and family are not in agreement, the question of removing life support will seldom arise, *Quinlan* notwithstanding. I would suggest that, since judicial supervision is unlikely to improve the accuracy of fact-finding and since there is ordinarily no major controversy among the parties involved, a judicial role is not essential to a "right" decision. Also, I would suggest that what is "right" in these cases is not objectively demonstrable anyhow [11, 16]. The only situation where I can see a clear need for judicial supervision is with respect to the patient who has no family willing or able to speak for him.

As the forums where these decisions are reached, hospitals may design their own rules or guidelines [17]. Through medical staff by-laws and regulations or provisions in intensive care procedure manuals, hospitals may require that attending physicians obtain consultations from qualified neurological specialists before recommending withholding or withdrawing care. And they might require that any joint decision by attending physicians and families to withhold or withdraw care be reviewed by an appropriate committee. The committee may appraise the clinical data and assure itself that the family has an adequate understanding of these data. Such a committee might also consider the legal overtones of a decision to withdraw or withhold life support, especially in those states (e.g., New Jersey, New York) where courts have addressed the issue. However, the question of a hospital's potential liability is best suited for discussions between the hospital's attorneys and its governing board. The committee would only be allowed to function if the governing board were satisfied that decisions to withhold or withdraw care are legally permissible in the hospital's locale.

Assuming courts are by-passed and no specific authorizing legislation exists, participants in decisions to withhold or withdraw care will have no prior assurance of immunity from civil or criminal liability. This lack of guaranteed immunity may constrain extra-judicial decisions, but this is not a necessary outcome. To be sure, the remote threat of liability encourages physicians to be exceptionally cautious in their appraisals of clinical data and to explore thoroughly with families the ethical aspects of a particular case [16]. Yet a deliberate and systematic inquiry which is conducted under the umbrella of standards formulated by a hospital should also promote confidence among physicians and families that the resulting decision is a proper one. If the inquiry uncovers uncertainties about prognosis or major ambivalences in a family's attitudes, it will become quite obvious that life-support measures should continue.

References

1. *Lane v Candura* (Mass. App., 1978). 376 NE 2d 1232.
2. *Mills v Rogers* (1982). 102 S Ct 990.
3. *Belchertown State School v Saikewicz* (Mass., 1977). 370 NE 2d 417.
4. *Matter of Storar* (N.Y., 1981). 420 NE 2d 64.
5. *In re Dinnerstein* (Mass. App., 1978). 380 NE 2d 134.
6. *Custody of a Minor* (Mass., 1982). 434 NE 2d 601.
7. *In re Quinlan* (N.J., 1976). 355 A 2d 647.
8. Jennett, B., and Plum, F. (1972). Persistent vegetative state after brain damage. *Lancet* **1**, 343–345.
9. *In re Quinlan* (N.J. Super., 1975). 348 A 2d 801.
10. Ryser, J. (1976). Every patient's situation is unique. *Am. Med. News* 11–13 (July 19).
11. Goldstein, J. (1977). Medical care for the child at risk: On state supervention of parental autonomy. *Yale Law J.* **86**, 645–670.
12. California Stat Ann, Health and Safety Code, Sec. 7185–7195 (1976). (West Cum. Supp.).
13. Beresford, H.R. (1977). The Quinlan decision: Problems and legislative alternatives. *Ann. Neurol.* **2**, 74–81.
14. Baron, C.R. (1979). Medical paternalism and the rule of law. *Am. J. Law Med.* **4**, 337–345.
15. Relman, A. (1978). The Saikewicz decision: Judges as physicians. *New. Engl. J. Med.* **298**, 499–500.
16. Burt, R.A. (1979). *Taking Care of Strangers: The Rule of Law in Doctor-Patient Relations.* The Free Press, New York.
17. Palmer, L.I. (1982). Dealing with terminally ill patients: An institutional approach. *Cornell Law Forum* **9**, 12–16.

Part II

Studies of the Biology and Neurology of Behavior: Implications for Ethics

Chapter 7

Out with the "Old" and in with the "New"—The Evolution and Refinement of Sociobiological Theory

ARTHUR L. CAPLAN*

Despite all that has been written about the subject of sociobiology during the nearly 10 years that have passed since the appearance of E. O. Wilson's justly canonized *Sociobiology: The New Synthesis* (1975), the subject is still beset by controversy and confusion. It is not at all clear exactly what the content is of sociobiological theory. Nor is it evident whether sociobiology constitutes a radical break with traditional inquiries into behavior in such fields as ethology and comparative psychology, or, is merely an extension of the models and approaches used in these fields. Nor is there yet much agreement over the implications of sociobiological theorizing for understanding human behavior in general, and, in particular, those moral and sociopolitical activities which are often cited (Sahlins, 1976) as paradigmatic examples of humankind's freedom from biological limits and constraints. Thus, this essay will have three major if somewhat immodest aims. First, the basic tenets of what I will refer to as the "old" and "new" sociobiology will be presented and discussed. Second, a critical assessment will be made of the "new" sociobiology. And, finally, some effort will be made to explain why sociobiological theorizing has been and continues to be a source of controversy by examining the implications of current sociobiological theorizing for morality.

*The Hastings Center, Hastings-on-the-Hudson, New York 10706.

What Exactly Is Sociobiology?

It is sometimes said that the allure of sociobiology to scientists consists in its ability to support conservative or bourgeois values (Allen, *et al.*, 1977). While the theory may hold this attraction for some scientists and surely holds it for at least a few nonscientists, the primary reason for sociobiology's ability to command scientific attention has little to do with values or politics. It has, instead, to do with two long-standing theoretical conundrums within evolutionary theory proper.

The problems posed by social behavior for Darwinism were first noted by its founder, Charles Darwin, in his *Origin of Species* (Darwin, 1859). It is often and repeatedly said (Peters, 1976) that Darwinian theory is tautologous—immune to any empirical counter-examples. But such a claim could only be made by someone who has not bothered to read the *Origin*. Darwin was keenly aware that his theory of evolution relied on three principles in explaining speciation and change in the organic world—the Principle of Inheritance, the Principle of Life's Dependency Upon Natural Resources, and the Principle of a Drive to Reproduce (Caplan, 1979). When these principles were supplemented with various empirical facts and generalizations, two conclusions could be deduced—that a struggle for existence would occur among living things, and that a natural process of selection would winnow out some organisms and favor the survival of others in the context of this struggle. Darwin had boldly asserted that all the variations of life found in nature could be explained by the invocation of the simple principles of reproduction, heredity, and dependency when supplemented with information concerning the prevailing environmental conditions.

Yet Darwin knew that two kinds of organic variation existed that were not consistent with his theory of evolution. Some highly varigated species existed in which certain members of a group did not reproduce. Soldier termites or drone bees were totally sterile and yet highly evolved. How, Darwin wondered (Darwin, 1859, 1871), could his theory, based on both reproduction and heredity, explain the existence from one generation to the next of creatures that did not reproduce and thus had no hereditary tendencies of any kind!

Similarly, Darwin, as a careful student of nature, was also well aware (Darwin, 1859) of the existence of various forms of altruism among members of the same and different species. Various birds and mammals, for example, sound warning cries when predators approach. Such cries would seem to benefit the group but to jeopardize the existence of the animal giving the warning. Darwin was puzzled over the prevalence and frequency of a trait, altruism, that seemed directly at odds with what should be expected to result from the process of natural selection in the context of a struggle for existence. Egoistical selfishness, not altruism, should emerge as the sole by-product of heredity, reproduction, scarcity, and struggle.

Darwin offered answers (Darwin, 1859) to both of these apparently contradictory cases, although he was not particularly enthusiastic about them. He argued that some traits, such as altruism or sterility, while deleterious to the individuals that possessed them, would, under some circumstances, benefit the group in which such individuals lived. However, Darwin was rather vague about the way in which group advantage could supercede individual advantage, and he wryly concluded his discussion of these two glaring counter-examples to his theory with the observation that an "over weening confidence" in his theory more than anything else made him believe that someday more light would be shed on these puzzling phenomena.

Various explanatory efforts were made to deal with the problems of sterility and altruism, but it was not until the publication in 1962 of V. C. Wynne-Edwards' book, *Animal Dispersion in Relation to Social Behavior* that a serious effort at a solution was proposed. Wynne-Edwards argued that selection for sterility in insect castes or for various forms of altruistic behavior could arise when animals were in danger of over-populating themselves out of existence. Wynne-Edwards' solution to Darwin's puzzles was to invoke the notion of group selection—that is, that under certain conditions of resource availability, population growth, and competition, groups which had tendencies toward altruism among their members would fare better than groups composed solely of rapidly reproducing egoists.

Wynne-Edwards' solution to the 100-year-old puzzle of altruism had great appeal. Indeed, its appeal was so great that it attracted the attention of numerous biologists, many of whom concluded that group selection, while a possible solution to at least one of the puzzles, was totally unacceptable on purely theoretical grounds (Williams, 1966).

Group selection depends upon the existence of mechanisms whereby traits advantageous to a group member can be passed along to future generations and spread to all group descendants. Darwinian theory, however, only postulates selection at the level of the individual—no means of group inheritance are acknowledged. Only traits beneficial or advantageous to specific individual group members can be passed on through reproduction. Thus, biologists such as W. D. Hamilton (1964) and G. C. Williams (1966), while recognizing the legitimacy of the problems posed by the original Darwinian puzzles, rejected Wynne-Edwards' solution as non-Darwinian.

The Emergence of the "Old" Sociobiology

The twin puzzles of evolution in sterile castes and altruism still demanded some sort of explanation in Darwinian terms. Over the 10 years that followed the rejection of Wynne-Edwards' group selection hypothesis, a set of solutions began to emerge. The efforts of such scholars as W. D. Hamilton, Robert Trivers, and Richard Alexander (Alexander, 1974; Hamilton, 1964, 1971; Trivers, 1971, 1972, 1974) to find a solution to these apparent counter-examples to Darwinian theory was summarized in E. O. Wilson's landmark volume, *Sociobiology: The New Synthesis* (Wilson, 1975). The three theoretical models developed during this period to explain both evolution in sterile castes and altruism; kin selection, reciprocal altruism, and parent–offspring conflict constitute what might be termed "old" sociobiology. Wilson's achievement in his highly influential book was to integrate these models with existing theories of population genetics, ecology, and demography to form what he termed "a new synthesis." What researchers in the late 1960s and early 1970s realized was that the puzzling nature of social behavior could be explained in Darwinian terms if the traditional unit of natural selection, the phenotype, was replaced or at best supplemented by the concept of the genotype as the unit of selection.

W. D. Hamilton (Hamilton, 1964, 1971) argued that since genes and genotypes were the units passed along from one generation to the next by reproduction, the genotype should be seen as the key locus of natural selection. He noted that from the point of view of natural selection similarities among genotype were indistinguishable. That is, if a particular genotype was favored in a given environment over other genotypes with respect to survival and reproduction, then it would tend to increase its representation in the gene pool of future generations. Thus, identical or similar genotypes will have the same survival and reproductive advantages no matter which organisms possess them.

Since, Hamilton observed, family members tended to have large numbers of identical genes, this being particularly so among haplodiploid insects, selection could favor a particular genotype and the traits resulting from it regardless of which particular individual possessed the genotype. Thus, selection, in favoring a particular genotype, would favor *all* individuals possessing that genotype.

The realization that selection acts on genotypes and not just phenotypes led Hamilton to see that both sterility and altruism could result as a consequence of Darwinian evolution. If selection were indifferent as to which particular individuals possessed a particularly advantageous genotype, then certain animals could help their kin either through direct aid or by not reproducing if, in so doing, they permitted greater numbers of similar genotypes to pass on to future generations. Thus, if, overall, more genotypes of a particularly favorable sort could be passed on by some members of a population remaining sterile and becoming specialized for food gathering or defense, then, under conditions of genetic identity or close similarity, sterility or altruism could appear and persisit.

Hamilton's theory became known as kin selection, and is one of the cornerstones of "old" sociobiology. However, his theory could only explain altruism among genetically related animals. Since there is a good deal of interspecific altruism in nature, some other additional means would have to be found to explain this behavior.

The two additional theoretical components supplied during the early 1970s were parental manipulation and reciprocal altruism. Richard Alexander and Robert Trivers (Alexander, 1974; Trivers, 1972; Trivers and Hare, 1976) noted that various forms of altruism exist among family members. However, certain conflicts of interest exist as well. For example, in many species it is much easier for males to produce sperm than it is for females to produce eggs. In many species, females bear a biological cost in carrying offspring and feeding them which males do not. These biological inequities should result in different reproductive strategies for males and females.

The realization that the "costs" of reproduction varied between the sexes in many species led sociobiologists (Hrdy, 1977; Symons, 1979) to realize that males and females would pursue different behavioral and reproductive strategies. For example, males might be more inclined toward promiscuity and less inclined toward child rearing, whereas the converse might be true for females. Also, females might be inclined to produce more males than females, depending on resource availability. These strategies do in fact appear to be manifest in the reproductive behaviors of many species and to be a source of cooperative behavior or conflict behavior, depending upon the kinds of environments and population distribution patterns that exist among various species.

The last component of the "old" sociobiology is the mechanism of reciprocal altruism. The best way of illustrating this mechanism is to imagine two individuals that each stand in some danger of drowning. Suppose that unaided, a particular individual has a 50% chance of drowning. However, the chance of any given organism drowning drops to about 5% if that individual is helped by another. Clearly, it would be to everyone's advantage in a group if each individual were prepared to help one another in situations where a real risk of drowning exists.

The basic point behind the idea of reciprocal altruism is to recast apparently altruistic behavior in terms of mutual benefits. What looks like altruism may actually be an example of reciprocal exchange between members of the same or different species. There is ample evidence that this kind of behavior does exist in nature (Barash, 1977; Chagnon and Irons, 1979; Wilson, 1975).

For example, among various fishes, some take the role of cleaners and other those who are cleaned. The little fish who do the cleaning get the advantage of protection from the large fish who are cleaned. And, rather than make a quick meal of the little fish, the larger fish benefits by having the smaller cleanse it of parasites and other debris. While it would appear as if the little fish in engaged in altruistic helping of the larger fish, what actually is going on is a paradigmatic example of reciprocal altruism wherein each participant in cooperative behavior gets something in return.

These three mechanisms of selection—kin selection, parental manipulation, and reciprocal altruism—were brought together by E. O. Wilson in his book, *Sociobiology: The New Synthesis* (1975). Wilson used these behaviors to explain a good deal of social behavior in the animal world. He also went on to speculate that some of these mechanisms might be at the heart of certain forms of social behavior among human beings.

It should be evident that all of these mechanisms imply a very tight link between genotype and behavior. Indeed, it was the postulation of tight control by genes over behavior that brought so much of sociobiology under critical attack in the late 1970s. Many critics felt (Wade, 1976) that, particularly with regard to human beings, the theory ignored the role of both culture and learning in the dissemination and prevalence of altruistic behavior among organisms. Sociobiologists (Wilson, 1978; Alexander, 1979) writing during the late 1970s did little to dispel this postulation of a close link between genes and behavior. Indeed, in some of their more speculative writings (Dawkins, 1976), they treated behavior as the direct consequence of particular genotypes, i.e., they postulated genes for the traits of homosexuality, aggression, parenting, sharing, etc. This sort of genetic determinism was at the heart of the initial negative reaction to sociobiology (Lewontin, 1977).

The Scientific Status of the "Old" Sociobiology

What kind of theory is "old" sociobiology? It might be thought that sociobiology represents a drastic break with evolutionary theory. The mechanisms of sociobiology might be viewed as an effort to cast aside outmoded Darwinian mechanisms of individualistic selection in favor of genotypic selection. If sociobiology met this description, one would expect a great deal of debate to have ensued among those defending individualistic interpretations of social behavior and those favoring genotypic views. However, with few exceptions, this did not take place. Many biologists (Gould, 1980) were critical of the adequacy of old sociobiology to explain all of social behavior, but few advocates of individualistic selection came forward to defend the adequacy of Darwinism itself.

Another way of looking at sociobiology is that it is simply an extension of Darwinism into the realm of behavior. One might see sociobiology as an effort by biologists to explain behavioral features of organisms by means of the same mechanisms that traditionally have been used in evolutionary biology to explain the physical features of creatures.

However, this view also seems false since, if one looks at the history of thought concerning the explanation of behavior, especially in the tradition of ethology, there are many, many examples of efforts by biologists to speculate about behavior and give explanations for various individual and social behaviors manifest in nature (Caplan, 1976, 1978, (1980b). It is historically false to say that evolutionary biologists prior to the appearance of the old sociobiology did not attempt to explain behavior by means of evolutionary theory.

A third possible view of old sociobiology is that the theory does not represent a drastic break with the older Darwinian theory of evolution, but does represent a change in the understanding of how the mechanisms of that theory operate. Evolutionary theory simply attempts to explain the evolution and persistence of various behaviors in terms of individual advantage. What makes sociobiology different is the realization that Darwinian theory must be amended to include the insight that selection can work not only at the level of the individual organism but also at the level of the genotype. On this view of sociobiological theory, the theory is not an extension of Darwinism to a new phenomena, behavior, nor is it a revolution within evolutionary biology itself. Rather, in invoking the mechanisms of sociobiology, scientists can be seen to be changing the causal level of interactions among the various forces at work in evolution.

I believe this view is the correct interpretation of sociobiology's relationship to Darwinism. The old sociobiology was an effort to shift the traditional forces and mechanisms of evolutionary theory to a new level of organizational complexity—the level of the gene. It does not represent a drastic break with older Darwinian theories, but rather a modification of existing theories with the result that certain problems are amenable to explanation using the same basic mechanisms and forces (Ruse, 1979).

It must be noted that scientific status of sociobiology was hardly the locus of the critical reaction that greeted the appearance of this view. Most of the reaction (Caplan, 1976; Caplan, 1978, 1980a,b; Wade, 1976) to sociobiology centered around normative and methodological issues rather than disputes over the status of the theory relative to earlier evolutionary ideas. The two primary sources of criticisms were ideological and methodological.

Many critics (Lewontin, 1977; Gould, 1980) felt that sociobiology was simply another version of genetic determinism. They noted that the theory posited such strong links between genotypes and behavior that it bore close resemblance to sexist and racist views that were simultaneously being propounded by many involved in I.Q. studies and in the inquiries into the biological basis of criminality. Sociobiology was held to be one more version of social Darwinism, the view that society must reflect nature if social arrangements are to be considered optimal. Thus, critics such as Stephen J. Gould, Richard Lewontin, and the Science for the People group (Allen *et al.*, 1977) argued that sociobiology was both racist and politically reactionary since it clung to a simplistic view of the causation of social behavior.

The same critics also argued that sociobiology as presented by Wilson was methodologically flawed since many of the basic models it utilized were untestable. They argued (Gould, 1980) that reciprocal altruism simply reinterpreted altruistic behavior in a way that was not amenable to empirical verification. It was too simple, according to the critics, to simply reinterpret true altruism in terms of reciprocal selfishness, and while this conceptual maneuver might be appealing, it would only be appealing to those who were ideologically disposed to see selfishness everywhere in nature. Critics of the old sociobiology dismissed it as just so much story-telling: speculation about possible ways in which behavior might have evolved was only that—speculation. Sociobiologists were merely spinning out tales, and these tales could not be subject to empirical verification, falsification, or test.

Ideology and Scientific Truths

It is certainly true that some of the speculations about human behavior current in the sociobiological literature in the late 1970s lent itself to easy ideological dismissal as crudely deterministic. Sociobiologists such as E. O. Wilson, Robert Trivers, Richard Dawkins, and others (Barash, 1977; Alexander, 1977, 1979; Symons, 1979) did in fact speculate in a free-wheeling manner about the possible biological bases of various complicated human and animal social activities. Moreover, there can be little doubt that the links between behavior and genes in most sociobiological writing of the time were opaque at best and simplistic at worst.

Nevertheless, it is important to realize that ideology is not always a good predictor of scientific bias in the explanation of behavior. If it is roughly true that biological explanations of behavior can be classified as either genetic or environmental, then history shows that persons of various political persuasions have held positions in both of these camps when it comes to explaining human behavior. For example, as some of Loren Graham's historical studies have revealed (Graham, 1981), one can find various Nazi scientists in the 1930s favoring environmental interpretations of human behavior, and, at the same time, find various Marxists strongly in favor of genetic and even eugenic interpretations of human behavior.

In one sense it is simply silly to argue that it is possible or desirable for scientists to attempt to dismiss the ideology and political views of their time. It simply is not possible for anyone to jump out of their own, time-bound, culturally mediated, ideological boots. What scientists believe, however, is that they have developed, over the years, a number of methods for testing various claims in order to have a partial check on the ideological and political biases that may be present in any particular scientific hypotheses.

What many of the critics of the old sociobiology failed to see was that a critical distinction has to be made between the role played by ideology and politics in the formation of scientific hypotheses and the role played by these factors in their assessment. It is a version of what philosophers term the "genetic fallacy" to argue that the validity or truth of a particular claim is suspect as a result of the source of origin of the claim. For example, while we may despise Adolph Hitler because of his racist and fascist beliefs, it is nevertheless true that Adolph Hitler could be the source of various true and valid claims about the world. In order to determine the truth or validity of his claims or anyone else's claims, we would have to subject them to various sorts of tests, falsifications, and evidence. We certainly want to be on guard against the use of ideology as a source of hypotheses in biology or in any other endeavor or inquiry. However, the mere fact that ideology makes its presence felt in the kinds of hypotheses that a particular scientist or a group of scientists are willing to advance does not mean that the entire notion of truth or validity evaporates from science. It is precisely as a protection against bias that scientists depend upon various methods of verification, test, and falsification for checking any and all claims made in the name of science.

Thus, it seems that many of the criticisms advanced by the Science for the People group and other Marxist-oriented critics of sociobiology in the late 1970s were flawed. While it might have been true that sociobiologists were influenced by the culture in which they lived in favor of explanations of altruism and social behavior that were based, in part, on selfishness, competition, greed, or manipulation, these facts do not in themselves impugn the possibility that the models of reciprocal altruism, kin selection, and parental manipulation are true or valid. The assessment of sociobiological models is quite separate from the issue of their origin within the mind of a particular scientist. And, as far as I know, no critic of sociobiology wanted to advance the view that all methods for assessing scientific hypotheses were suspect since they too were ideologically soiled. The fact remains that the standards of evidence and test prevalent in science today have stood the test of time over many centuries, and there is little reason to think that they are so distorted by ideological values as to be useless.

The Sociobiology Debate—Round 1

The first round of the sociobiology debate seemed to end with the clear advantage being retained by proponents of the theory. While the critics made some headway in arguing that sociobiological accounts of the causation of behavior were too crude and too deterministic, for the most part sociobiological explanations of the puzzles of altruism and sterile castes seemed to appeal to the majority of evolutionary biologists. Moreover, many scholars in anthropology, psychology, sociology, and even in the humanities (Midgley, 1978; Singer, 1981; Stent, 1980) found themselves attracted to some of the models put forward under the banner of the old sociobiology.

A large number of books and collections appeared in which various methodological and ideological issues were debated. E. O. Wilson himself entered the fray with a book, *On Human Nature* (Wilson, 1978), which inquired into the meaning of sociobiology for understanding human nature. Wilson argued that sociobiology indicated that certain utopian hopes of social thinkers about the malleability of man had to be rejected in favor of more realistic understandings of the limits and constraints that existed as a consequence of human nature. Moreover, Wilson went on to argue that the kind of scientific materialism implicit in sociobiological interpretations of human social behavior might serve as a substitute for religious and spiritual interpretations and explanations of various social customs and mores.

The claims of critics of old sociobiology to the effect that the models of this view—kin selection, reciprocal altruism, and parental manipulation—were untestable were, for the most part, ignored. Indeed, a large number of scientists set out to examine through empirical field research the adequacy of these models for explaining various forms of social behavior in a large number of species. In particular, studies were done (Alexander and Tinkle, 1981) to determine the genetic relationships that prevailed within various populations to see if kin selection could be at the heart of certain forms of social behavior in birds, insects, and higher mammals. Many anthropologists (Chagnon and Irons, 1979) also attempted to see whether the social patterns found in various cultural groups of human beings were amenable to explanation by the models of sociobiology. At least in some cases (Hrdy, 1977) the models were vindicated by empirical studies.

Thus, the course of the debate up until the end of the 1970s was that various claims were made on behalf of sociobiology, with some having far more empirical substance than others. These claims were met with a flurry of ideological and methodological criticisms which, in turn, led at least some scientists to put sociobiological models to the test of empirical evidence. The most devastating criticism offered of sociobiological accounts of social behavior, that sociobiology was far too crude in its understanding of the complexities that existed between selection and genotype, remained unanswered.

The "New" Sociobiology

The second stage in the sociobiology debate might usefully be dated from the appearance in early 1981 of *Genes, Minds and Cultures*, which was jointly written by Charles Lumsden and E. O. Wilson (Lumsden and Wilson, 1981). This book surely must be seen as an effort to respond to the central criticism of the older sociobiological research program concerning genetic determinism. The most trenchant of all the criticisms mounted against the old sociobiology was that the link between genes and social behavior was far too tenuous, reticulate, and complex to make the sociobiological approach worthwhile in the study of the behavior of human beings or any other advanced species.

Lumsden and Wilson attempted to meet this criticism in their new book. They admitted that social behavior in humans and other species is best understood as the end product of a host of complicated interactions among genes, genotypes, phenotypes, and cultural factors. Moreover, they admitted that earlier allusions within the sociobiological literature to the causal links between genes and behavior by metaphors such as that genes "kept culture on a leash" were not sufficient as accurate depictions or explanations of the genesis of social behavior.

Lumsden and Wilson attempt to provide in *Genes, Minds and Cultures* a comprehensive theory of gene culture "co-evolution." Briefly, what they propose to do is to analyze culture into units they refer to as culturgens. Culturgens are defined as:

> A relatively homogeneous set of artifacts, behaviors, or mentifacts [mental constructs having little or no direct correspondence with reality] that either share without exception one or more attribute states selected for their functional importance or at least share a consistently recurrent range of attribute states within a given polythetic set. (Lumsden and Wilson, 1981)

Culturgens, in this definition, are introduced into the analysis of gene culture coevolution to serve as the operational units of culture. They are intended to be the cultural equivalents of the physical traits that were so long studied, investigated, and manipulated by biologists who were interested in organic phenotypes. The kinds of culturgens that illustrate the clusters suggested in the Lumsden and Wilson definition fall into such categories as incest avoidance, village fissioning, and the depth of decolletage in women's dresses. These are the kinds of cultural traits which Lumsden and Wilson proposed to analyze with their new theory.

The pathway that Lumsden and Wilson posit from genes to culture is a rather complicated one. Basically, they argue that genes and genotypes work together through what they term "epigenetic rules" to determine certain basic phenotypes and properties among populations of organisms. For example, genotypes place restrictions on the kinds of sensory perception that different species are able to engage in. While there are certainly variations in the perceptual abilities and skills of the individual members of any given species, the fact remains that these differences are quite small relative to the possible range of sensory skills and attributes that a creature might have had, had it not inherited genetic programming from its ancestors.

Lumsden and Wilson then go on to argue that these primary epigenetic rules and the limitations they impose upon organisms result in further limits on the kinds of customs, mores, and behaviors that can arise in any given society. For example, organisms endowed with particular types of nerves, sensory organs, and brain patterns can only learn particular types of behaviors and social arrangements. Their idea is that culture is in part determined by the ways in which the primary epigenetic rules of development interact with the environment to produce creatures that are more or less conditionable or open to the acquisition of various customs and behaviors. They term the range of behaviors compatible with any particular array of organic and neuro-logical features "ethnographic curves." It is these ethnographic curves which determine the cultural inheritance of a particular society, be it animal or human. It is also these ethnographic curves which are confronted by various selection forces in the environment.

On the Lumsden and Wilson theory of gene culture coevolution, the pathway from particular gene to particular cultural manifestation is mediated by a number of levels of causal interaction. Selection cannot look through directly to genotypes, but must act upon behavioral variations which are under the control of secondary rules and learning functions. These are, in turn, under the control of primary epigenetic rules and genotypes. It is this long story from genes to culture that has to be invoked if we are to understand the ways in which natural selection acts upon behavior.

It is important to realize in looking at this newer version of sociobiological theory that Lumsden and Wilson have taken their critics quite seriously and have attempted to meet the objection of simplicity by posing a theory of gene culture interaction that is far more complicated than any that appeared in the original or older writings of sociobiology. Moreover, it should also be evident that this theory is still quite consistent with the basic tenets of Darwinism. As was the case with the older sociobiology, the newer sociobiology basically admits that there are many levels of interaction between the environment and the organism—genotypic, developmental, phenotypic, environmental, and cultural. The theory acknowledges the complexity of the inter-actions, but shows that certain universals of behavior will inevitably result when the proper combinations of genotype, development, environment, and culture are in place. Like the old sociobiology, the new sociobiology make a concession toward causal complexity, not by overthrowing Darwinism, but, instead, by acknowledging the fact that many levels of causal interaction exist in nature, and, that not all of these have been adquately reflected in older versions of evolutionary theory or sociobiology.

Methodological Issues Confronting the "New" Sociobiology

Perhaps the most serious methodological issue confronting the "new" sociobiology is the issue of how culturgens are to be defined. It is not clear that culture can be divided into units or even continuously varying sets in the same way that organic traits are individuated and classified. It is simply not clear that culturgens must be particular units.

For Darwinian evolution to occur, the units of heredity must remain constant and relatively immune to environmental alteration if selection is to pick and choose among genotypes. As Darwin's original critics noted nearly a century ago, if interbreeding results in the blending of hereditary materials, the efficacy of the mechanisms of natural selection would be greatly diminished. What is not clear is why we ought to assume that culture ought to be divided into units akin to those we are familiar with from the study of genotypes and phenotypes. There are many reasons for suspecting that culturally based habits, customs, behaviors, and personal ideas are malleable entities open to both directional and nondirectional environmental influences. It is not even clear that cognitive states, such as desires, wishes, intentions, and beliefs, can be divided into units by means of space–time frameworks. The kinds of conceptual frameworks that we use to chart the physical world may simply not be of help in trying to chart the cultural or mental world (Caplan, 1982b).

Moreover, it is not at all clear what sorts of behaviors ought to be counted as significant in terms of culturgen identification. Why should we use the kinds of behaviors and customs that seem perceivable to us? There is no real independent evidence adduced in Lumsden and Wilson's book for using skirt length, village fissioning, incest avoidance, or any other common behaviors as basic or important in the analysis of culture. Since we can only "see" culturgens, using our own skewed epigenetically determined modes of perception, why should we believe that these tools will give us a true picture of the way the cultural world actually is? Without some way of independently verifying the utility and accuracy of culturgenic units, it seems that there is all too much room for bias and ideology to enter into the classification of culture in ways that are not amenable to the standard methods of scientific assessment and test.

The current critics of the "new" sociobiology were quick to unleash the same sorts of criticisms against it as had been leveled against the older sociobiology. The theory of gene-culture coevolution was dismissed as reductionistic, materialistic, and untestable (Caplan, 1982a; Leach, 1981). Moreover, many critics repeated the old argument that political biases of Lumsden and Wilson had again been allowed to infest the analyses that they offer of social behavior.

It must be admitted that there is some basis for worry about the role played by bias in the selection of particular customs and behaviors as illustrative of the culturgens posited by Lumsden and Wilson. Nevertheless, the same standards of assessment invoked to check value biases alleged about the old sociobiology remain apt in examining the same issue regarding the new sociobiology. I happen to believe that the units selected as culturgenic are suspect as a result of the lack of independent evidence for the selection of these units as *the* units of culture. Particularly important in this respect is the absence of independent laws in which the units suggested by Lumsden and Wilson play key roles as either the causes or the effects of human interactions. But rather than dwell on this particular issue, I would like to conclude this paper with some general comments about the broader ethical implications and possible misapplications of the old and new socio-biologies.

Sociobiology and Ethics—Reciprocal Altruism at Last?

It has to be understood that the primary reaction among those in philosophy, religion, political theory, and other fields within the humanities to sociobiology old and new has been dread combined with anger. Humanists are even more resentful than their social scientific kin of the territorial claims made by sociobiology. Sociobiologists have been quite explicit in their belief that the humanities are in acute need of biological assistance. Those in the humanities have reacted with amusement and outrage to such claims. After all, those in the humanities are familiar with a long history of efforts by scientists to scientize ethics and other areas of the humanities (Caplan, 1976). Sociobiology appears to many humanists to be one more example of scientists pounding their fists upon the table of reductionism and threatening to "bury" any and all humanists who stand in their way!

There are some very practical reasons why those in the humanities do not believe that sociobiology has much to offer by way of insight into ethics. First, few humanists understand sociobiological theory. There are simply few people in many of the humanities who are in a position to understand or appreciate current theoretical developments in this field. Second, it is also true that many humanists are deeply suspicious of the scientific standing of sociobiology and evolutionary theory as well. However, as some of the arguments in this paper should make evident, the worries of those in the humanities about the tautologous or circular nature of sociobiological theory are not well grounded.

I believe that it is imperative to overcome the long-standing suspicions of those in ethics and the humanities toward any efforts to introduce scientific theorizing into the study of morality. It is true that there are difficulties in drawing normative conclusions from factual statements. Simply being told that particular dispositions or tendencies exist among human beings toward certain forms of behavior does not prove that these behaviors are either desirable or permissible. Nevertheless, it is also reasonable to assume that constructing moral norms and rules which make demands on human beings that go beyond their physical, emotional, or psychological capacities is an exercise in frustration at best. The notion that morality should not exceed biological and cultural capacity is a truism that makes it evident that those in ethics would be well advised to pay close attention to claims about limits and constraints on human action as they are presented by those in the biological sciences if ethics is to remain consistent with biological possibilities. It simply makes no sense to talk about ethical ideals that are beyond the reach of human conduct, motivation, and emotion.

One way in which the kinds of biological claims made by both the old and the new sociobiology seems to be particularly germaine to moral theory is in the study of political theory. Most of contemporary political theory still reaches back to certain myths about human nature that have been inherited from 16th, 17th, and 18th century political theorists (Caplan, 1980b). Older thinkers such as Hobbes, Rousseau, Adam Smith, and John Locke all were very much influenced by the anthropological and sociological views of their day in formulating their ideas about politics and social organization. Unfortunately, very little of the empirical information available to these thinkers at the time they wrote is consistent with current understanding about human behavior and human social activities. Thus, the time seems ripe for a break from traditional ways of grounding political ideas by means of mythical, "social contracts," or states of nature, and, to integrate current hypotheses in behavioral biology into political thinking.

Another way in which sociobiology is surely relevant to ethics is in the examination of the evolution of ethics itself (Singer, 1981). It is surely hard to study the evolution of behaviors as complex as sociality, and, no doubt, many sociobiologists have made terrible gaffs in talking about the evolutionary history of morality among early human beings, but, nevertheless, descriptions concerning the origins of early forms of ethical behavior in human beings and the nature of early normative rules and principles are vital to understanding contemporary ethics. While those in ethics like to believe that their theories are insulated from the social and biological past, the fact remains that this is no more true for ethical theories than it is for scientific theories. Unless we know how our thinking about ethics has evolved historically, it will be difficult to know what is bias and what is valid in morality.

The final way in which sociobiology and other biological theories would seem to be directly relevant to the study of ethics is in the examination of moral psychology. Sociobiology, in both its old and its new guises, makes various claims about the nature of human motivation. Are people naturally inclined toward selfishness or altruism? Is it possible to reinterpret apparent cases of truly altruistic human behavior in terms of egoistic behavior? If we expect people to come to the aid of each other in various kinds of situations, we must believe that this is psychologically attractive for them, or that we can design various types of social rewards in order to encourage such behavior. Sociobiologists clearly want to define altruism and egoism solely in terms of the consequences of various behaviors—they tend to focus on the outcomes of behavior for reproductive fitness. Philosophers, however, look more frequently to motives.

Consider the alternative interpretations given by an ethicist and a biologist to the standard sociobiological example of rescuing someone in danger of drowning. From the sociobiological point of view, someone who puts themselves at risk to save another must be doing so either as a consequence of kin selection considerations, or from reciprocal altruism. The ethicist, however, assesses the situation by inquiring into the rescuer's motives. If you act from love or a consideration of others, you are altruistic—consequences be damned! For biologists, failed attempts to rescue others hardly count in the evolutionary scheme of things. But for those in ethics, the outcome of a rescue effort is much less important than the fact that it was undertaken in the first place. Here we encounter one of the most basic conceptual gulfs that separates the biological sciences from morality—the difference between reason and causes, motives and consequences. Those in biology and ethics must address together the conceptual question involved in assessing and interpreting human action if the true normative implications of sociobiology are to be properly understood.

There are many reasons for remaining cautious about assigning too much weight to speculations in the new sociobiology about the causes of human social behavior. As Lumsden and Wilson themselves suggest, the link between genetic endowment and cultural outcome is long, reticulate, and complex. Moreover, sociobiology is a relatively new endeavor and many of its models will undoubtedly fail to stand the test of empirical adequacy and nomological utility. However, ethics and political theory seem to have remained far too isolated from contemporary work in the biological sciences, and the time appears fortuitous for that disciplinary isolationism to finally be cracked. If questions of values can be moved away from worries about ideology toward more sustained critiques of the normative and meta-ethical claims made by various persons in light of sociobiology, then a fruitful form of conceptual reciprocal altruism would appear to be on the intellectual horizon.

References

Alexander, R. D. (1974). The evolution of social behavior. *Ann. Rev. Ecol. Systematics* **5**, 325–384.

Alexander, R. D. (1977). Evolution, human behavior, and determinism. In:

P.S.A. *1976* Suppe, F., and Asquith, P., eds.). pp. 3–21. P.S.A., East Lansing, Michigan.

Alexander, R. D. (1979). *Darwinism and Human Affairs*. University of Washington Press, Seattle.

Alexander, R. D., and Tinkle, D. W. (eds.) (1981). *Natural Selection and Social Behavior*. Chiron Press, Ann Arbor, Michigan.

Allen, E., *et al.* (1977). Sociobiology: The new biological determinism. In: *Biology as a Social Weapon* (Sociobiology Study Group, eds.). Burgess, Minneapolis, Minnesota.

Barash, D. P. (1977). *Sociobiology and Behavior*. Elsevier, New York.

Caplan, A. L. (1976). Evolution, ethics and the milk of human kindness. *Hastings Center Rep.* **6** (2) (April 1976), 20–25.

Caplan, A. L. (1978). *The Sociobiology Debate*. Harper & Row, New York.

Caplan, A. L. (1979). Darwinism and deductivist models of theory structure. *Studies Hist. Philosophy Sci.* **10** (4) (December 1979), 341–353.

Caplan, A. L. (1980a). A critical examination of current sociobiological theory. In: *Sociobiology: Beyond Nature/Nurture* (Barlowe, G. P., and Silverberg, J., eds.), pp. 97–121. Westview Press, Boulder, Colorado.

Caplan, A. L. (1980b). Of mice and men: Of the human sciences and the humanities. *Hastings Center Rep.* **10, 6**, (December 1980), 38–40.

Caplan, A. L. (1982a). Say it just ain't so: Adaptational stories and sociobiological explanations of social behavior. *Philosophical Forum* **13, 3** (Spring, 1982), 144–160.

Caplan, A. L. (1982b). Stalking the wild culturgen. *Behavioral Brain Sci.* **5, 1** (March 1982), 8–9.

Chagnon, N. A., and Irons, W. P. (eds.) (1979). *Evolutionary Biology and Human Social Behavior*. Duxbury Press, North Situate, Massachusetts.

Darwin, C. (1859). *On the Origin of Species*. A facsimile of the first edition with an introdution by Ernst Mayr, published in 1967. Harvard University Press, Cambridge, Massachusetts.

Darwin, C. (1871). *The Descent of Man and Selection in Relation to Sex*. Appleton, New York.

Dawkins, R. (1976). *The Selfish Gene*. Oxford University Press, New York.

Gould, S. J. (1980). Sociobiology and the theory of natural selection. In: *Sociobiology: Beyond Nature/Nurture?* (Barlowe, G. W., and Silverberg, J., eds.), pp. 257–269. Westview Press, Boulder, Colorado.

Graham, L. (1981). *Between Science and Values*. Columbia University Press, New York.

Hamilton, W. D. (1964). The genetical evolution of social behavior. *J. Theoret. Biol.* **7**, 1–52.

Hamilton, W. D. (1971). Geometry for the selfish herd. *J. Theoret. Biol.* **31**, 295–311.

Hrdy, S. (1977). *The Langurs of Abu*. Harvard University Press, Cambridge, Massachusetts.

Leach, E. (1981). Biology and social science: Wedding or rape? *Nature (London)* **291** (May 1981), 267–268.

Lewontin, R. C. (1977). Sociobiology—A caricature of darwinism. In: *P.S.A.*

1976 Suppe, F., and Asquith, T., eds.), Vol. 2. P.S.A., East Lansing, Michigan.

Lumsden, C. J., and Wilson, E. O. (1981). *Genes, Mind and Culture.* Harvard University Press, Cambridge, Massachusetts.

Midgley, M., *Beast and Man.* Cornell University Press, Ithaca, New York, 1978.

Peters, R. H. (1976). Tautology and evolution in ecology. *Am. Naturalist* **110**, 1–12.

Ruse, M. (1979). *Sociobiology: Sense or Nonsense?* D. Reidel, Boston, Massachusetts.

Sahlins, M. (1976). *The Use and Abuse of Biology.* University of Michigan Press, Ann Arbor, Michigan.

Singer, P. (1981). *The Expanding Circle.* Farrar, Straus and Giroux, New York.

Stent, G. S. (ed.) (1980). *Morality as a Biological Phenomena.* University of California Press, Berkeley, California.

Symons, D. (1979). *The Evolution of Human Sexuality.* Oxford University Press, New York.

Trivers, R. L. (1971). The evolution of reciprocal altruism. *Quart. Rev. Biol.* **46**, 35–57.

Trivers, R. L. (1972). Parental investment and sexual selection. In: *Sexual Selection and the Descent of Man, 1871–1971* (Campbell, B., ed.). Aldine, Chicago, Illinois.

Trivers, R. L. (1974). Parent-offspring conflict. *Am. Zoologist* **14**, 249–264.

Trivers, R. L., and Hare, H. (1976). Haplodiploidy and the evolution of social insects. *Science* **191**, 249–263.

Wade, N. (1976). Sociobiology: Troubled birth for a new discipline. *Science* **191**, 1151–1155.

Williams, G. C. (1966). *Adaptation and Natural Selection.* Princeton University Press, Princeton, New Jersey.

Wilson, E. O. (1975). *Sociobiology: The New Synthesis.* Harvard University Press, Cambridge, Massachusetts.

Wilson, E. O. (1978). *On Human Nature.* Harvard University Press, Cambridge, Massachusetts.

Wynne-Edwards, V. C. (1962). *Animal Dispersion in Relation to Social Behavior.* Oliver and Boyd, Edinburgh.

Chapter 8

Some Paradoxical Goals of Cells and Organisms: The Role of the MHC

Jerram L. Brown*

Introduction

A biological perspective on morals may not solve personal dilemmas, but it may clarify their roots. In this overview, I shall survey some parallels between social cells and social organisms with respect to the genetic and behavioral basis of cooperation and conflict. No lessons are preached here about how individuals or cells should behave. In particular, I do not connect any moral stance with genetics. On the contrary, I would like to suggest that moral dilemmas are more frequently caused by genetic factors than solved by them.

The historical foundation for this essay was laid by W. D. Hamilton (1964) in one of the most brilliant contributions to evolutionary biology of the 20th century. The classical theory of natural selection had since Darwin's time accepted the idea that aid by parents to their descendant kin—namely, their offspring and their descendants—was selected because of a net genetic advantage to the parents. That is, on average parents were more than repaid for their risks and energetic expenses while caring for their young. Hamilton simply extended this generalization to include nondescendant kin in addition to descendant kin— hence the name kin selection given by Maynard Smith (1964). This

*Department of Biological Sciences, State University of New York, Albany, New York 12222.

possibility had been realized by earlier authors (e.g., Fisher, 1930); Hamilton, however, both first developed the mathematical theory for the extension and first pointed out to incredulous biologists its far-reaching implications. Hamilton's theory enabled the formulation of many new hypotheses concerning the evolution of all types of social interactions among animals, but its particular relevance to the worker castes of the social insects was immediately apparent. Thus, it is not surprising that it was an entomologist, E. O. Wilson (1975), who most effectively urged extension of inclusive fitness theory to mankind. In theory then, an animal, X, can influence the frequency of its genes in future generations in two ways: (1) *directly*, by effects on kin that received genes of X via gametes of X; and (2) *indirectly*, by effects on kin who received some of the same genes not from X, but from a common relative of X (Brown, 1980; Brown and Brown, 1981).

The most common situation for indirect effects occurs when a worker ant, bee, wasp, or termite aids the reproduction of its mother or sister. In many social insects, the workers never reproduce and they achieve their genetic purpose in life solely by the indirect route. If it is assumed or demonstrated that a worker forfeited an opportunity to utilize the direct route (have its own offspring), then such behavior in the terminology of evolutionary biology constitutes altruism. To demonstrate empirically and convincingly that a particular species-typical behavior is altruistic is nearly impossible (Brown, 1978; Brown and Brown, 1981; but see Metcalf and Whitt, 1977; Noonan, 1981). The direct loss in altruism can in theory be compensated genetically via indirect effects.

The terms altruism and kin selection continue to be misused even in reputable textbooks and especially in the mass media. By Hamilton's definition, care of one's own offspring is not altruism, since parental care normally has a net direct beneficial effect. Parental care is, however, correctly said to be kin selected.

Hamilton's theory of inclusive fitness caused renewed interest in the ability of kin to recognize each other and react in ways that might be predicted by the theory. Beneficial effects of a behavior on kin do not require an ability to discriminate kin for the behavior to be selected, so long as these effects are restricted to contexts in which kin are likely to be found together. Nevertheless, an ability to discriminate kin from nonkin would enhance the beneficial effects of kin selection. The present essay explores speculatively some possible genetic mechanisms for kin recognition between cells and between individuals and attempts to place them in a broader perspective than has heretofore been appreciated.

Special attention is given in this essay to the major histocompatibility complex (MHC) and related gene families. For readers who are unfamiliar with MHC, I recommend the one-page "beginner's guide to MHC function" by Matzinger and Zamoyska (1982). Useful summaries of MHC genetics and evolution are provided by Klein *et al.* (1981) and Robertson (1982).

Two related paradoxes are explored, one behavioral and one genetic. Cooperation and competition—the sources of morals and ethics—are universal in living systems. Cooperation and competition occur at all structural levels, from DNA base sequences to communities of species. Our first paradox is that at every structural level cooperation is competitive. More specifically, cooperation at one level is a device to compete at the next higher level, with benefits reflected back to the cooperating level.

For example, people cooperate as members of a team—so that their team may compete more effectively—with benefits to themselves. Identical twins are very good at cooperation with each other. Similarly, genes of an organism may be said to cooperate as a team that we call an individual. An individual is more than a gene's way of making more genes; an individual represents a genuine cooperative effort made by its genetic team in competition with other such teams for limited resources. Such situations—embodying multilevel competition and cooperation in a nonbenign environment—generate patterns of behavior at the level of the cell and at the level of the individual.

The rules that guide cooperative and competitive behaviors have their roots in two seemingly contradictory "goals" of living matter. These are:

1. The pursuit of genetic similarity—for the sake of cooperation—between cells and between organisms.
2. The pursuit of genetic dissimilarity—for the sake of cooperation—between genes.

The second and more fundamental paradox is rooted in the organization of the genome of the individual. This is that the same genetic, cellular, and behavioral mechanism may be employed for these different goals. I shall examine each of these two goals, first conventionally at the cell level, and then at the individual level.

The Level of the Cell

Genetic Similarity

Genetic similarity is strongly promoted at the cell level by mitosis, which results in an organism composed of cells that are genetically nearly identical. Not all cells in an organism, however, come from the same stock. Some are invaders: viruses, bacteria, protozoa, and other parasites. The biological need to discriminate against foreign cells and viruses has been a principal concern of decades of research in immunology. Defense of the genetic integrity of the individual requires the existence of mechanisms to recognize self as distinct from nonself. In the vertebrates, these mechanisms have been investigated by transplanting tissues from one animal to another. Such experiments have gradually elucidated the nature of the genetic control of self-recognition and some of the cellular interactions involved. In this process T-killer cells and T-helper cells play prominent roles.

Genetic control of graft rejection does not reside uniformly through-out the genome. In each species it is concentrated in one region. This is true for tunicates (Scofield et al., 1982), which are protochordates, for mice (Mus musculus), where the region is referred to as the H-2 system (Klein et al., 1981), and for man, where the region is referred to as the HLA system (for human lymphocyte antigen). Collectively the verte-brate system (excluding the protochordates) has been called the major histocompatibility complex, commonly referred to as the MHC (Klein, 1975). The MHC appears to be a major force for promoting cooper-ation among cells of an individual and for the recognition and killing of foreign cells and viruses. In the mouse, the MHC is located on the 17th chromosome; in man, on the 6th chromosome. Structurally similar sequences are also found elsewhere in the genome, and evidence for a history of duplications, deletions, and pseudo-genes is rapidly accumu-lating (Robertson, 1982).

Genetic Dissimilarity

Goal 2, the pursuit of genetic dissimilarity between uniting gametes for the sake of improved gene teams, is mediated by the machinery of genetic recombination, sexual reproduction, and mating behavior. Meiosis, of course, is fundamental to this entire complex of related phenomena, starting with a genetically uniform population of cells, called an organism, meiosis provides a genetically diverse population of gametes.

In the process of fertilization of the egg by the sperm, cell fusion takes place. It is important that a gamete avoid fusing with another that is genetically identical to itself. The goal of sexual reproduction is to enable fertilization between gametes that are genetically dissimilar. Many elaborate mechanisms prevent self-fertilization in multicellular organisms, but all of them were apparently preceded phylogenetically by a more basic and universal mechanism—cell recognition at the membrane surface, as found in the immune system. It has been suggested "that the cell–cell communication system has evolved from the membrane structures originally meant for cell recognition in the mating process" (Monroy and Rosati, 1979, p. 166). This insight connects today's immune system in the vertebrates with a phylogenetically primitive sexual function. As we shall see below, remnants of this functional connection may still persist.

A genetic mechanism to facilitate dissimilarity between uniting gametes has been described for the tunicate *Botryllus* (Scofield *et al.*, 1982). This example has special interest because tunicates are protochordates—closely related to the ancestors of all vertebrates—and because this genetic region has MHC-like properties. *Botryllus* eggs appear to "resist fertilization by sperm from the same colony." This property derives from the MHC-like "fusibility locus," which also controls graft rejection. In short, a genetic mechanism controlling cooperation among similar cells also influences the sexual union of gametes—favoring genetic similarity in the first case and genetic dissimilarity in the second.

The connection between sexuality and the MHC may be closer than is commonly realized. A prominent theory for the origin of sexuality holds that its primitive and perhaps still principal function is to enhance the ability of long-lived hosts to combat adaptations by disease agents with short life cycles, such as viruses and bacteria (Hamilton, 1980). Since it is the MHC that carries much of this responsibility, the MHC may be the principal genetic reason for sexuality in the vertebrates.

The Level of the Individual

Genetic Dissimilarity

The advent of multicellular individuals enabled more complicated mechanisms to participate in seeking the goal of genetic dissimilarity between uniting gametes. Genetic self-incompatibility mechanisms prevent selfing in hermaphroditic plants. Gonochorism—the phenomenon of separate sexes—prohibits selfing even more effectively in most animals.

With the advent of mobility by both sexes, animals were freed from reliance solely on membrane mechanisms to prevent close inbreeding. Furthermore, most if not all, mobile animals have been provided with mechanisms of behavioral selectivity in choice of a mate.

Mate Choice, Imprinting, and Phenotypic Matching

Lack of selectivity in choice of a mate can have disastrous consequences if carried to either extreme. On the one hand, biological
isolating mechanisms largely prevent the production of abnormal
offspring that often result from extreme outbreeding—hybridization
between species, for example. On the other hand, a collection of rather
poorly known epigenetic effects tends to hinder the production of
defective offspring that may result from the opposite extreme, close
inbreeding. There may be an optimum between these extremes, though
it may not always be clearly defined.

Between these extremes exist many important parameters and
mechanisms of choice. One of these mechanisms involves comparison
of a potential mate to a reference phenotype that is typically learned
early in life and is correlated with the individual's own genotype. After
sexual imprinting, as this learning process is sometimes called,
individuals address courtship to phenotypes like those with which they
associated when young, but not identical to them in every detail
(Lorenz, 1935). In other words, animals may learn to prefer a
phenotype like their parent or rearing associates but they tend to avoid
mating with parents or sibs. This means that in mating, animals are
taking into consideration not only obvious species characteristics but
also subtle details that enable the avoidance of close inbreeding.

For example, Japanese quail (*Coturnix coturnix*) of both sexes reared
only with sibs tend to associate as adults with members of the opposite
sex who are unfamiliar first cousins, in preference to unfamiliar birds
more distantly related, and in preference to the sibs with whom they
had been reared (Bateson, 1982). Similarly, female mice preferred the
smell of unrelated males of the same strain to males who were their
brothers or males of another strain (Gilder and Slater, 1978).

These results suggest that animals in paying attention to phenotypic
details of potential mates might be influenced in their choice by genetic
effects on the reference phenotype. Paradoxically, these genetic
influences may be provided by the environment—namely, from the
mother, as experiments have shown. If this were true, mating preferences could be influenced by genetically determined environmental
influences that could be used to minimize dangers of close inbreeding
or the dangers of deleterious homozygosity at specific loci.

MHC Effects in Mus

What evidence is there of such genetic effects? The basic discovery was
reported in a study of the MHC by K. Yamizaki, E. A. Boyse, V. Mike,
H. T. Thaler, B. J. Mathieson, J. Abbott, J. Boyse, Z. A. Zayas, and

L. Thomas of the Sloan–Kettering Cancer Center in 1976. Males of a given MHC type were given a choice of females differing genetically from each other only at an MHC locus and having the same genetic background. In the majority of combinations tested, males favored females dissimilar to their own MHC type. It appears that male mice can discriminate among females on the basis of MHC-determined odors and that these may influence the sexual preferences of males, tending to favor matings between mice dissimilar at certain MHC loci. Tests on F_2 segregants reconfirmed earlier findings, provided further controls, and provided evidence for MHC preferences by females (Yamazaki et al., 1978).

As these authors emphasize, the MHC, because of its phenomenal genetic variability, probably has a greater capacity to express phenotypic individuality than any other part of the genome. The scientific breakthrough represented in the above work lies in the demonstration that an "identity card" (Dausset, 1981) whose use was previously thought to be restricted to cell-level interactions can be used at the individual level for biologically important goals such as facilitating genetic dissimilarity among uniting gametes.

Several problems remain:

1. How important is the MHC relative to the remainder of the genome in such behaviors?
2. Are mating preferences determined entirely by genetically correlated imprinting effects or are olfactory receptors also coded in a lock and key manner?
3. If the receptors are involved, are their properties coded by the same loci that determine olfactant diversity or by different loci?
4. Finally, do any of these laboratory phenomena have relevance to natural populations?

In my opinion, the present evidence is inconclusive on these basic questions, and honest differences of opinion exist (compare Bekoff (1981) with Andrews and Boyse (1978) and Yamazaki et al. (1978) with regard to the matter of receptor genes).

MHC Effects in Man

Do any of these phenomena have significance for man—a species in which culture frequently overrides biology—often with detrimental effects? It would be premature to give a simple answer; but there are strong hints that the subject is worth pursuing.

That impressive genetic variability is found at MHC loci in man is well known. That MHC genes affect the fertility of human marriages and the physical well-being of their offspring is not so well known. Spontaneous abortions, difficulty in bearing embryos to term, and neural tube defects of babies appear in higher frequency than usual when father and mother have certain MHC genes in common (Komlos et al., 1977; Schachter et al., 1979a,b; Takeuchi, 1980; Patillo, 1980; Beer et al., 1981).

Close inbreeding is generally deleterious in mammals, but the extent to which MHC genes are responsible for inbreeding depression remains to be clarified. People can recognize their relatives and discriminate among them by odor. They can also distinguish the sexes and identify their own odor and that of their mate (Porter and Moore, 1981; and included references). In short, even in man, surprising variability in odor and ability to discriminate odors are present, and there are biological penalties for not using them to avoid close inbreeding. Culture and other mechanisms may, of course, also be used.

The immune system probably has even greater significance in our lives than the above remarks suggest. Autoimmune effects are becoming more widely appreciated; and speculation has linked them to language disorders, various diseases, spatial and verbal cognition, handedness, levels of sex hormones, sexual differentiation, and the ontogeny of brain assymetries (Marx, 1982; Goodfellow and Andrews, 1982). The family of genes to which MHC belongs is thought to have far-reaching effects on embryonic development (Artzt and Bennett, 1975).

Genetic Similarity

The pursuit of genetic similarity for the sake of cooperation is best exemplified at the cell level—by the multicellular organism. The same goal also can be recognized, however, at the level of the individual. Just as cells have a problem with genetically dissimilar intruders, so too do individuals—at least in certain circumstances.

Clonal Attached Colonies

Cooperation among individuals, according to inclusive fitness theory (Hamilton, 1964), should be greatest among members of a clone, in which individuals are genetically identical. Clonal cooperation is most reliable genetically when the members never lose contact with each other—much as the cells in multicellular organisms. This is most dramatically shown in the "colonial" coelenterates, in which polyps of a "colony" bud off and differentiate but remain attached to a common base—the hydroids, corals, and siphonophores being good examples. In the tunicate, *Botryllum*, which grows on a similar principle, graft rejection between organisms has been shown to depend on an MHC-like locus (Scofield *et al.*, 1982).

Clonal Unattached Individuals

Unlike most coelenterates, the sea anemones (class Anthozoa) grow as separate polyps rather than as a group of mutually attached polyps. This seems to discourage cooperation and role differentiation among individuals except in one species of this group, *Anthopleura elegantissima* (Francis 1973a,b, 1976). These animals live attached to rocks in the ocean. Reproduction can be by fission, leading to a colony of clonal descent. These colonies in nature consist of closely packed members of the same sex. Between colonies are uninhabited areas. When members of different clones touch each other, there is an electrical impulse followed by aggressive behavior (Lubbock, 1980). Tentacles are extended and nematocysts are discharged into the "enemy." In the center of such clones, individuals are specialized for reproduction, with large gonads. On the periphery, individuals are specialized for defense, with many tentacles specialized for stinging neighbors. This behavior does not occur between members of the same clone.

Defender morphs among mobile members of a clone have also been described for aphids (Aoki, 1977) and for a polyembryonic encyrtid parasitoid wasp (Cruz, 1981) for interspecific use. Among nonclonal species, morphological castes, including defender morphs are known in the eusocial insects and in one mammal, the naked mole rat (*Heterocephalus glaber*) (Jarvis, 1981). Kin recognition has been described in some of these cases.

Paradox Resolved

In this essay, I have discussed two paradoxes—each one fundamental to the organization of living matter. The first, that cooperation is competitive, sets the stage for conflicts of interest between the individual and its group. This conflict may be found at every level of structural organization, though I have here discussed it only at the levels of cell and individual.

Cooperation among cells, including differentiation, is effective apparently mainly when the cells are genetically identical, though exceptions can be made to occur by grafting (Lubbock and Allbut, 1981). Cooperation with morphological differentiation between individuals is also most effective when the individuals are genetically identical. Genetic similarity, however, is apparently not enough; proximity and utility are also required. Proximity means that individuals have frequent opportunity to interact socially. It is favored by structural attachment and by family social structure. Utility means that genes causing individuals to interact, through cooperation and role differentiation, have an ecological advantage because of which they increase in frequency. In short, cooperation and role differentiation are favored by three factors: genetic identity, proximity, and utility.

The second paradox results from the first. In this case, it is the genes that compete and the genes that cooperate in order to compete. Though an isolated cell or individual may prosper, isolated genes do not. Teamwork among genes is, therefore, essential; but how are the teams formed?

Team formation occurs by recombination followed by fertilization. For genes to survive through numerous cycles of recombination and fertilization they must have "chosen" their teammates well. One of the factors important in the gene team is heterozygosity. Heterozygosity tends to confer numerous advantages on zygotes ranging from disease resistance to social dominance. To facilitate adaptive heterozygosity the behaviors of cells and of organisms are adapted to seek genetic dissimilarity between uniting gametes. Our second paradox, then, is this: Adaptations favoring genetic similarity among cooperating units exist compatibly in the same organisms that favor genetic dissimilarity at fertilization.

How does this happen? I argue here that one mechanism—probably present in all organisms—serves both of these conflicting goals. The mechanism is the comparison of phenotypes (phenotypic matching), particularly those that correlate highly with genotypes. By this mechanism both genetic similarity and dissimilarity can be detected and evaluated. The conflicting goals of pursuit of genetic similarity and dissimilarity can both be served by this mechanism.

If comparison of phenotypes were all there were to this story, I would not be telling it. Many authors have made this point. Recent work on the genetics of cell and individual interactions suggests a further simplifying theme. To be useful in achieving our paradoxical goals, the method of phenotypic comparison requires genetic individuality. Only if there is sufficient genetic individuality are these goals possible and worth pursuing. With respect to genetic individuality, all parts of the genome are not equal. Certain regions are much more variable than others. These hypervariable regions are expected to be the most critical in determining and in assessing genetic individuality.

Hypervariable regions have been discovered in the entire range of chordates from protochordates to man. Their various functions appear to be mediated by the control of cell-surface proteins. These membrane structures are important in the identification of dissimilar gametes and dissimilar invading cells in the diploid organism. Membrane structure apparently triggers defensive reactions in sea anemone clones. Membrane structures controlled by the MHC somehow determine olfactory individuality, which in turn affects the behavior of individual mice. I propose that it is not too much to expect that MHC-controlled membrane features also influence properties of tissues and of organs. In this way genes controlling membrane properties might have effects on other sensory modalities. It is at least conceivable that hypervariable regions of the genome might have phenotypic correlates that could be seen, heard, or felt, as well as correlates that can be smelled. It is known, for example, that other loci in the same gene family as MHC and tightly linked to MHC have a fundamental bearing on development (Artzt and Bennett, 1975).

As research continues on the molecules and cell biology of development, evolutionary biologists should be alert for implications that might help us to understand the abilities of individual coelenterates, Japanese quail, sweat bees, mice and men to discriminate subtle differences in genetic individuality.

Morals and Genetic Individuality

Most questions of morals or ethics can be traced to the conflict between the goals of individuals as individuals and the goals of groups to which these individuals belong. The existence of groups, their size, frequency, permanence, and power, all derive in part from the extent to which group goals coincide with individual goals. Morals, ethics and group norms are codified by groups and become guidelines for their members. As the common goals of a membership change, so too do the group guidelines or norms change.

With respect to changing human moral and ethical positions, my contribution, if any, is to remind us all that:

1. Cooperation and competition are paradoxically inextricably intertwined and engendered by the same environmental forces.
2. The nature of cooperation and of competition derives also from genetic diversity and genetic individuality.

3. Genetic individuality is derived from the basic processes of meiosis, recombination, and fertilization.
4. Cells and organisms simultaneously:
   (a) seek genetic similarity for cooperation, under favorable conditions of proximity and utility; and
   (b) seek modest genetic dissimilarity in partners for their gametes.
5. Hypervariable regions of the genome are the prime candidates for:
   (a) the product of genetic individuality, and
   (b) the provision of mechanisms with which to evaluate individuality; and finally
6. One such hypervariable region, the MHC, has already been shown to have:
   (a) many of the hypothesized qualities at the cell level, and
   (b) some, at least, of the hypothesized qualities at the individual level.

In conclusion, I would like to urge cooperation among molecular biologists, cell biologists, immunogeneticists, animal behaviorists, developmental psychologists, ecologists, and anthropologists in exploring the fundamental paradoxes of life.

Acknowledgments

I would like to thank L. Flaherty, W. D. Hamilton, C. Barkan, and E. Brown for discussion of some of the ideas in this paper. This research is sponsored by the National Science Foundation.

References

Andrews, P. W., and Boyse, E. A. (1978). Mapping of an H-2 linked gene that influences mating preference in mice. *Immunogenetics* **6**, 265–268.
Aoki, S. (1977). *Colophina clematis* (Homoptera:Pemphigidae), an aphid species with soldiers. *Kontyu, Tokyo* **45**, 276–282.
Artzt, K., and Bennett, D. (1975). Analogies between embryonic (T/t) antigens and adult major histocompatibility (H-2) antigens. *Nature (London)* **256**, 545–547.
Bateson, P. (1982). Preferences for cousins in Japanese quail. *Nature (London)* **295**, 236–237.
Beer, A. E., Gagnon, M., and Quebbeman, J. F. (1981). Immunologically

induced reproductive disorders. In: *Endocrinology of Human Infertility: New Aspects* (Crosignani, P. G., and Rubin, B. L., eds.). Academic Press, London.

Bekoff, M. (1981). Mammalian sibling interactions. In: *Parental Care in Mammals* (Gubernick, D. J., and Klopfer, P. H., eds.). Plenum, New York.

Brown, J. L. (1978). Avian communal breeding systems. *Annu. Rev. Ecol. Syst.* **9**, 123–155.

Brown, J. L. (1980). Fitness in complex avian social systems. In: *Evolution of Social Behavior: Hypotheses and Empirical Tests* (Markl, H., ed.). Verlag Chemie, Weinheim.

Brown, J. L., and Brown, E. R. (1981). Kin selection and individual selection in babblers. In: *Natural Selection and Social Behavior: Recent Results and New Theory* (Alexander, R. D., and Tinkle, D., eds.). Chiron, New York.

Cruz, Y. P. (1981). A sterile defender morph in a polyembryonic hymenopterous parasite. *Nature (London)* **294**, 446–447.

Dausset, J. (1981). The major histocompatibility complex in man. *Science* **213**, 1469–1474.

Fisher, R. A. (1930). The genetical theory of natural selection. Clarendon Press, Oxford.

Francis, L. (1973a). Clone specific segregation in the sea anemone *Anthopleura elegantissima. Biol. Bull.* **144**, 64–72.

Francis, L. (1973b). Intraspecific aggression and its effect on the distribution of *Anthopleura elegantissima* and some related sea anemones. *Biol. Bull.* **144**, 73–92.

Francis, L. (1976). Social organization within clones of the sea anemone *Anthopleura elegantissima. Biol. Bull.* **150**, 361–376.

Gilder, P. M., and Slater, P. J. B. (1978). Interest of mice in conspecific male odours is influenced by degree of kinship. *Nature (London)* **274**, 364–365.

Goodfellow, P. N., and Andrews, P. W. (1982). Sexual differentiation and H-Y antigen. *Nature (London),* **295**, 11–13.

Hamilton, W. D. (1964). The genetical evolution of social behavior. I and II. *J. Theor. Biol.* **7**, 1–51.

Hamilton, W. D. (1980). Sex versus non-sex versus parasite. *Oikos* **35**, 282–290.

Jarvis, J. U. M. (1981). Eusociality in a mammal: Cooperative breeding in naked mole-rat colonies. *Science* **212**, 571–573.

Klein, J. (1975). *Biology of the Mouse Histocompatibility-2 Complex.* Springer, New York.

Klein, J., Juretic, A., Baxevanis, C. M., and Nagy, Z. G. (1981). The traditional and a new version of the mouse H-2 complex. *Nature (London)* **291**, 455–460.

Komlos, L., Zamir, R., Joshua, H., and Halbrecht, I. (1977). Common HLA antigens in couples with repeated abortions. *Clin. Immunol. Immunopathol.* **7**, 330–335.

Lorenz, K. (1935). Der Kumpan in der Umwelt des Vogels. *J. Ornithol.* **83**, 137–213, 289–413.

Lubbock, R. (1980). Clone-specific cellular recognition in a sea anemone. *Proc. Nat. Acad. Sci. USA* **77**, 6667–6669.

Lubbock, R., and Allbut, C. (1981). The sea anemone *Actinia equina* tolerates allogenetic juveniles but alters their phenotype. *Nature (London)* **293**, 474–475.

Marx, J. L. (1982). Autoimmunity in left-handers. *Science* **217**, 141–144.

Matzinger, P., and Zamoyska, R. (1982). A beginner's guide to major histocompatibility complex function. *Nature (London)* **297**, 628.

Maynard Smith, J. (1964). Group selection and kin selection. *Nature (London)* **201**, 1145–1147.

Metcalf, R. A., and Whitt, G. S. (1977). Relative inclusive fitness in the social wasp *Polistes metricus*. *Behav. Ecol. Sociobiol.* **2**, 353–360.

Monroy, A., and Rosati, F. (1979). The evolution of the cell-cell recognition system. *Nature (London)* **278**, 165–166.

Noonan, K. M. (1981). Individual strategies of inclusive-fitness-maximizing in *Polistes fuscatus* foundresses. In: *Natural Selection and Social Behavior* (Alexander, R. D., and Tinkle, D., eds). Chiron Press, New York.

Patillo, R. (1980). Histocompatibility antigens in pregnancy, abortions, infertility, preclampsia, and trophoblast neoplasms. *Am. J. Reprod. Immunol.* **1**, 29–34.

Porter, R. H., and Moore, J. D. (1981). Human kin recognition by olfactory cues. *Physiol. Behav.* **27**, 493–495.

Robertson, M. (1982). The evolutionary past of the major histocompatibility complex and the future of cellular immunology. *Nature (London)* **297**, 629–632.

Schachter, B., Muir, A., Gyves, M., and Tasin, M. (1979a). HLA-A,B compatibility in parents of offspring with neural-tube defects or couples experiencing involuntary fetal wastage. *Lancet* (14 April), 796–799.

Schachter, B., *et al.* (1979b). HLA compatibility and phenotype in anencephaly and recurrent fetal loss. *Am. Assoc. Clin. Histoc. Testing.* Seventh Annual Meeting, Orlando, Florida (abstract).

Scofield, V. L., Schlumpberger, J. M., West, L. A., and Weissman, I. L. (1982). Protochordate allorecognition is controlled by a MHC-like gene system. *Nature (London)* **295**, 499–502.

Takeuchi, S. (1980). Immunology of spontaneous abortion and hylatidiform mole. *Am. J. Reprod. Immun.* **1**, 23–28.

Waldman, B. (1981). Sibling recognition in toad tadpoles: The role of experience. *Z. Tierpsychol.* **56**, 341–358.

Wilson, E. O. (1975). *Sociobiology*. Harvard University Press, Cambridge, Massachusetts.

Yamazaki, K., Boyse, E. A., Mike, V., Thaler, H. T., Mathieson, B. J., Abbott, J., Boyse, J., Zayas, Z. A., and Thomas, L. (1976). Control of mating preferences in mice by genes in the major histocompatibility complex. *J. Exp. Med.* **144**, 1324–1335.

Yamazaki, K., Yamaguchi, M., Andrews, P. W., Peake, B., and Boyse, E. A. (1978). Mating preferences of F2 segregants of crosses between MHC-congenic mouse strains. *Immunogenetics* **6**, 253–259.

Chapter 9

Motives and Metaphors in Considerations of Animal Nature

Colin Beer*

Introduction

"What proof is there that brutes are other than a superior race of marionettes, which eat without pleasure, cry without pain, desire nothing, know nothing, and only simulate intelligence as a bee simulates a mathematician?" (Huxley, 1893; quoted in Boden, 1972, p. 281). If we could persuade ourselves that animals are really inanimate, we should have little if any reason to think that there are ethical issues raised by their use in research in the neural and behavioral sciences. Although some people get emotionally attached to objects such as cars, or houses, or paintings, they usually stop short of claiming rights for them. We do, of course, value such things; and pleas for the preservation of historic buildings, works of art, natural habitats, and so forth, usually include appeal to a sense of responsibility or some such moral consideration. Yet what is usually argued in these cases is that *people* would be the poorer for loss of whatever is in jeopardy; unless one is an animist, the things themselves are not credited with existing to themselves in some way that entitles them to be considered as having interests, points of view, rights, which should be taken into account when there is a question about how they are to be treated. By and large, whether we regard a thing or a creature as having a claim on

*Institute of Animal Behavior, Rutgers University, Newark, New Jersey, 07102.

ethical consideration will depend upon the extent to which it appears to be sentient in the ways in which we are—possessed of sensations, feelings, emotions, motives, intentions, and so forth.

There are philosophical aspects to this matter, which have been argued at least since the days of Descartes, and can still provoke perplexity. I shall say a little about them, as I think that they bear on our theme. I shall be especially concerned about ways in which language affects and reflects the conceptions we have of the creatures with which we deal. My position is a bit like that of the old lady whom E. M. Forster reported as saying: "How can I know what I think till I see what I say" (Auden, 1968, p. 22).

Other Minds

When I was a child, I sometimes used to wonder whether what other people called "blue" was the same as what I called "blue." I was bothered by the thought that there may be no way for me to know for sure. Even though I and my friends agreed about which things were blue, there seemed to be the possibility that the property that these things had in common might appear different to my friends from what it did to me. Since the experiment of occupying someone else's mind, of being them and looking at the world through their eyes, was presumably ruled out, I concluded that I should have to go through life forever uncertain about this question.

Later I came to see the question as part of the broader issue that philosophers refer to as the problem of "other minds." Whether by the route of Cartesian systematic doubt, or by that of Humean skepticism, one can, I found, arrive at a solipsist position from which the existence of everything apart from one's present awareness is called into question. In particular, since mine is the only mental life I can ever experience, how can I be sure that there are any others? Since one cannot always trust what people say, perhaps one never should. Perhaps all the talk, gesture, expression of feeling, and so on, which constitute social intercourse, are mere facade, behind which there is nothing apart from what one projects from one's own experience, and expectation derived from it.

Descartes' sharp dichotomy between mental and material substance, and his doctrine that animals, in contrast to humans, partake only of the material, meant that animals are wholly body and hence incapable of thought or feeling. (Recall "Hugh Miller": "What is mind? No matter! What is matter? Never mind!") Since, according to this view, bodies are merely machines, all that might seem to us to be expression of reason, or emotion, or any kind of sentient life in animal behavior, must really be nothing other than surface by-product of the working of automatic mechanism, no more significant of mental content than the squeaking of a pully or the flicker of a flame. One would like to think that it was this belief that made it tolerable for vivisectionists such as Magendie to get on with the job even when their victims were squirming and screaming in what might otherwise be taken as pain; but there is reason to suspect, at least in the case of Magendie, that the opposite was true, that the apparent infliction of pain was part of what made the work interesting (see Taylor, 1963).

No matter how cogent the arguments for solipsism might be, we find it impossible to live according to their conclusions. The philosopher who has just proved, Q.E.D., that no mind exists but his own, will, in the next instant, reprimand his pupils for inattention, have a fit of temper because his secretary has failed to bring him his tea on time, and remind himself to take home flowers for his wife so that she will go on thinking that he loves her. Sane human life is lived in an unquestioned world of other selves, whose words we generally take at face value, whose feelings we generally try not to hurt, whose states of mind we generally think we understand by bringing sympathy, empathy, sensitivity to bear on how they look and move, laugh and cry. To be sure, we are sometimes taken in by lies and faked feeling, and we sometimes misunderstand what we are told and misinterpret what we hear in a sigh or see in a face; but such lapses are possible only against a background of a shared system of signs, in exchanging the tokens of which honesty and trust can generally be taken for granted.

However, the degree to which the system of signs is shared between people varies according to whether they speak the same or different languages, come from the same or different social classes, religions, cultures, or races. Foreigners seem to jabber (whence "barbarians"); to Europeans orientals are inscrutable; during wars the alien characteristics of the enemy can all too easily be made to appear as those of creatures less than human in their beastliness. But beasts also lie on the scale of intelligibility of expression: chimpanzees seem scarcely less than human in much of their social interaction; our dogs and cats and horses persuade us by looks and whines and nudges much in the way that our children do; as we pass to birds and fishes, to crabs and insects, to worms and sponges, the signs of sentience fade or become increasingly opaque. Griffin (1976) appealed to this graded continuity in signs of mentality in his discussion of "the question of animal awareness." Even Descartes, in his later writings, qualified his conception of the animal as machine to the extent of conceding that at least some animals are subject to sensation, and so have consciousness of sorts. For reasons like those that make it impossible to take solipsism seriously in real life, we find it hard to draw a margin to mentality between human kind and the rest of the animal kingdom.

Language Games

Philosophers have found more to bring against solipsism than the practical impossiblity of living as though it were true. Wittgenstein, for example, made a case against the possibility of the private language, which solipsism implies. If my way of assigning meaning to a word consisted in ostensive naming of something known only by introspection, not only would I have no idea what anyone else meant by it; I should not even myself be sure of the meaning of the word. To know the meaning of a word is to know how to use it correctly; but where there is no possibility of checking whether the word is used correctly or incorrectly—whether, for instance, this color that I now call "blue" is the same as that to which I first privately pinned the word—then the distinction evaporates; at best I can have only the impression that I am following my rule (Wittgenstein, 1958). Wittgenstein's conclusion was that a language is necessarily public, and its words work by having "outward criteria."

The private language argument was only part of Wittgenstein's treatment of the problem of the meaning of "meaning." In his first major work, the *Tractatus Logico-philosophicus* (1922), he took the position that all nontautologous sentences serve the one task of describing sense-experience. Similarly the logical positivists had taken the statement of empirical fact as the type for all meaningful statements. In the *Philosophical Investigations* (1958), however, Wittgenstein rejected this view in favor of one which developed Bertrand Russell's point that grammar can be a false guide to logic. Indeed he recognized a far richer diversity of logical type than Russell had envisaged, and he shifted attention from meaning to use: "Don't ask for the meaning, ask for the use," became the new edict; and in harmony with it was the slogan: "Every statement has its own logic" (Urmson, 1956, p. 179). Describing the world is only one of many uses to which we put language, and even when we use it for description we do so in ways which differ from one another and are not all reducible to one type. This liberal attitude, according to which each philosophical problem needs to be approached on its own terms, rather than via a preconceived program, was expounded by Wittgenstein in his talk about "language games." Just as playing-cards must be used in conformity with different rules in different games, so words and sentences have to be understood as subject to different patterns of logical relationship in different contexts, if philosophical muddle is to be avoided or cleared up. For example, to argue, as many have done, that all our actions are governed by self-interest, is to flout the rules of the language game in which the notion of self-interest plays its part, and so render it vacuous, like playing tennis without a ball, or playing poker by yourself. If nothing can be allowed to be unselfish, in spite of appearances to the contrary, then the proposition that all action is motivated by self-interest becomes analytic, immune to test, and hence of no explanatory or predictive value; and so we lose the notion of selfishness as well. Another example: Compare "What do you have to do to turn on the radio" and "What do you have to do to lift your arm?" (asked of a normal person). In spite of the similarity in grammatical form, the second question raises a kind of perplexity different from the first; what could one say in reply other than something like: "I just lift it, that's all!" for there is nothing further or more elementary on our side of the action for the notion of doing to engage. Yet, of course, there is much else that goes on—nerves transmit, muscles contract, and all that; but description at that level is in another language game in which the notion of doing and the notion of self that it entails do not play.

Metaphors and Motives

There are many things that we say of ourselves that we also say of animals. Can we always, or ever, assume that the rules of the language game are the same for both cases? There is no problem with terms like "fat" and "dead," "running" and "living," "being burned" and "getting poisoned." For example, I can say that Sydney Greenstreet was fat, and that Charlotte the pig was fat, without raising any philosophical hares. But what about terms like "selfish" and "cruel," "searching" and "wanting," "being intimidated" and "getting punished?" If, for example, I say of an animal that it is an utterly selfish brute, do I imply the possibility that it has alternative courses of conduct open to it, as is implied by Darcy's saying, in *Pride and Prejudice*: "I have been a selfish being all my life, in practice though not in principle"? The emergence of operational definition in animal psychology reflects how tough-minded behavioral scientists considered such terms to be applicable to animals only if shorn of much of the connotation they carry when used in human contexts.

However, operational definition is too blunt an instrument for the finer points of logical diversity, as Charles Taylor demonstrated in his book *The Explanation of Behaviour* (1964). For example, it cannot adequately handle the class of idioms characterized by what philosophers, following Brentano, call "intentionality." "Intention," in the sense of conscious purpose governing action, is only one of these intentional idioms, which also include wanting, imagining, knowing, believing, thinking, perceiving, and a host of other mental acts and states. They have in common that they are all about or directed toward a content, which Brentano called the intentional object, and in a manner that Brentano believed to be peculiar to mentality (see Boden, 1972; Dennett, 1969; Chisholm 1967). Thus to want is to want something; to think is to think of something; to believe is to have a belief about something; you cannot simply want, or think, or believe *simpliciter*, in the way in which you can twitch, or cough, or age. Brentano's claim about mentality is not universally conceded; but the intentional idioms have one or more logical characteristics which set them apart from other idioms with which they can be compared. "Referential opacity" is one of these characteristics. If I say that I know Neil Armstrong, it does not necessarily follow that I should also claim to know the first man to walk on the moon, for I may be ignorant of the fact that they are one and the same. Referential opacity obtains when different ways of picking out the same particular cannot always be substituted for one another without affecting the truth of the statement in question. Thus Hamlet intended to kill the man behind the arras, but he did not intend to kill Polonius. Compare this to "Polonius was Ophelia's father." Since Polonius was the man behind the arras, it follows that "The man behind the arras was Ophelia's father" is also true. Statements like this latter, which are about things or events or relations in the physical world, conform to extensional logic, in which substitution of codesignating terms does preserve truth value. Another characteristic differentiating some intentional sentences from extensional sentences is exemplified by comparison of the following: "John wants an ice cream" and "John licks an ice cream." Whereas John's wanting an ice cream does not imply the existence of any particular ice cream, his licking of one does. When names or descriptions are used in intentional sentences there need be no implication that they refer to anything that actually exists (except as an intentional object), in contrast to extensional sentences, which do imply the existence of that to which they refer, at least as a general rule. There are counter examples which pose problems when these and other logical characteristics are claimed as criteria of intentionality (see Chisholm, 1967; Urmson and Cohen, 1968; Grice, 1957, 1969; Dennett, 1969). Nevertheless the point has been made that the logic of language shows greater diversity than used to be supposed or is even now generally recognized, the differences being especially marked between what can roughly be categorized as the psychological and the physical.

What of the use of psychological or intentional language in talk about animals? Does such use carry with it the same logic that applies in contexts of human thought and action? There are certainly cases where this is not so. When, for example, a biologist says that the eye-like markings on the sides of a caterpillar or the wings of a butterfly are a means of bluffing or deceiving predatory birds, he does not imply, and is not taken to imply, that the caterpillar or butterfly deliberately put the markings there in order to bluff or deceive. He would imply and be taken to imply such intention if he were reporting that an escaped convict bluffed his way to freedom by posing as a prison guard. When Richard Dawkins (1976) writes of selfish genes, he clearly does not intend us to think that the molecules can be held morally responsible for the way they carry on.

Dawkins's characterization of the genes as selfish, like a great deal of the talk in sociobiology, is obviously metaphorical. Metaphor can be more than mere poetic decoration on what might be better put in plain prose. Max Black (1962) has eloquently argued for the heuristic value of metaphor in science, and discussed cases where the metaphorical mode is so sustained and systematic in application to a scientific matter that is serves to construct an implicit conceptual framework—part of the paradigm, as Thomas Kuhn would say.

The "motive for metaphor" has often manifested itself in biology. The most famous case must be Darwin's extension of the thought and concepts of *laissez-faire* economics from the Victorian world of commerce to the evolving world of nature. In recent evolutionary ethology and sociobiology, we again find a pervasive fabric of metaphor representing natural process in terms of human policy. There is such a deal of traffic in costs and benefits, trade-off and pay-off, investment and self-interest, that someone straying in off the street could well go out with the impression that the talk was of economics rather than biology. The frequency of reference to selfishness, manipulation, strategy, deceit, honesty, altruism, game-playing, and so forth, could give the impression that the student of behavior assumes the animals to be rational, calculating, ethical, scheming beings who live according to conventional codes of conduct. Even for those who work in this area of study, to hear such people as Richard Dawkins and Amotz Zahavi in full cry is to be carried along by the pack of figures to such effect that it is easy to forget that they do not literally mean what they say.

However, we are seldom tricked by the tropes for long. Either reflection reveals the case to be such that only a metaphorical sense can be intended, or the author tells us that this is so. For instance, in reference to a bird's use of a warning signal in a deceptive way, Dawkins writes: "We would not mean he had deliberately intended consciously to deceive" (1976, p. 68). The value of using analogies drawn from the field of human thought and action to make points about animal social behavior is plainly seen in the new questions that have been posed in terms of them, such as the questions about altruism, "evolutionary stable strategies" (Maynard Smith, 1974, 1976), communication as manipulation (Dawkins and Krebs, 1978), and in the way in which these questions have germinated a mushrooming of evolutionary, ecological, and genetic studies. Indeed this growth is so rampant as to cause some uneasiness about the extent to which it draws off interest in other kinds of question.

The swings of fashion have their returns, in science as elsewhere, so we can expect that this dominating preoccupation with how features of animal social behavior can be given evolutionary interpretation will be a passing thing. In the meantime, however, other important questions suffer neglect, or, perhaps even worse, are regarded as somehow absorbed into or settled by the evolutionary interest. Dawkins (1976, p. 66) marks some of the distinctions; for example: "The details of the embryonic developmental process, interesting as they may be, are irrelevant to evolutionary considerations." Irrelevance makes the mind grow blinder, especially when, as here, the issue is far from clear. Elsewhere Dawkins is less careful about keeping unconnected considerations apart. In spite of claiming: "I am not concerned here with the psychology of motives" (Dawkins, 1976, p. 4), he frequently says things like " . . . gene selfishness will usually give rise to selfishness in individual behaviour" (Dawkins, 1976, p. 2), which sound very like motivational conclusions drawn from a metaphorical premise about molecules.

Mary Midgley (1978, 1979) has taken Dawkins and other sociobiologists to task for appearing to argue in this fashion. She pleads that metaphorically talking as though genes had motives is no substitute for a motivational psychology of the individual, and should not be made to appear so. She is quite abusive—to my taste excessively so—in dealing with Dawkins, whom she represents as confirming George Eliot's observation that we all get our thoughts entangled in metaphors and act fatally on the strength of them: "Dawkins brings in gene motivation because his account of individual motivation is a total failure; in fact he switches from one to the other with bewildering speed every time he gets into difficulty" (Midgley, 1979, p. 446). Midgley thus, in contrast to Max Black, emphasizes that argument by analogy is a species of fallacy.

Dawkins can be read as anticipating this line of criticism. By defining selfishness, altruism, and so forth in "behavioural terms," that is to say in terms of reproductive consequences, he dispenses with the intentional connotations, and can consistently claim that his talk is not about motives at all, if he uses the words consistently in accordance with his definitions. But this is a bit like the left hand taking away what the right has given. Without their intentional connotations, these words lose what gives them most of their figurative force in this context. Is the racy language of *The Selfish Gene* really a cover for something quite staid?

I doubt whether the behavioral or biological redefinition of intentional terms can be sustained any more consistently than the behavioristic attempts could be. Much of the sociobiological use of such terms is suggestive of "double-think." Intensional logic is so primitive to words like lying, deceit, self-interest, altruism and so on, that most of the instances of their use make little clear sense unless this logic is assumed.

Let us take reciprocal altruism, for example. Trivers (1971) introduced this concept to sociobiology to try to account for cases of acting in another's interest at cost or risk to oneself, where genetic relationship to the other individual is too remote for kin selection to be operative. The idea is that if I help you when you are in distress today, I can count on you to do likewise should you find me in distress tomorrow. The long-term benefits of such mutual aid are supposed sufficiently to outweigh the costs as to give the altruistic system a selective edge over alternatives. But the altruistic system is clearly vulnerable to invasion by cheats. Someone who accepts help but never gives any will enjoy all the benefits of the system without incurring any of the costs, so cheats should prosper and spread, at least until the ratio of cheats to altruists arrives at the equilibrium of an evolutionary stable strategy. For such equilibrium to be possible, the cost of cheating must go up with increase in the proportion of cheats, and the most obvious way for this to come about would be for altruists to refuse help to individuals they come to know to be cheats. Such discrimination against cheats would entail keeping a record of who helps and who does not, hence individual recognition and memory, both of which are intentional capacities (recognizing and remembering are both referentially opaque). Trivers therefore predicted that reciprocal altruism will arise only where individual recognition is possible, and group composition sufficiently stable and long-lasting for there to be repeated opportunities for giving and taking. To date, as far as I know, the only well-documented case of reciprocal altruism, apart from its human manifestations, is from a study of baboons (Packer, 1977). Humphrey (1976) has speculated that the origins of intellect may lie in the mental capacities required by reciprocal altruism. In any event, the example illustrates how easily the talk can go from a selectionist conception of altruism to a motivational or intentional one.

Manipulation—Adaptive Consequence or Motivated Action?

The rarity of reciprocal altruism, and its apparent restriction to animals close to the human case, both phyletically and comparatively, might be taken as supporting the view that ascription of mentality should be reserved for people, or animals behaving very like people, and hence that the use of intentional idioms in other contexts should be understood as excluding intentional implications. Again, this is easier said than done, but there is also some doubt about whether it should be done in all cases. There are occasions when one might question whether the rule against intentional ascription should be followed, cases where an intentional idiom might be thought to apply more or less literally, rather than in the Pickwickian sense of sociobiological positivism.

When Dawkins and Krebs (1978) wrote of communication as manipulation, they explicitly stated that they were adopting a "cynical gene" view of social interaction, according to which signal use is to be understood in terms of how it serves to promote replication of the genes underlying it, by getting other individuals to cooperate, capitulate, acquiesce, be appeased, be submissive, and so forth, to that end. This conception of manipulation draws no distinction between a flower attracting a bee by its color, a bitch in heat who mindlessly makes the dogs come running in response to her smell, and a political campaigner who mindfully gets people to vote for him by the eloquence of his rhetoric. But the distinction between the mindless and the mindful can often be drawn, and there are many cases of animal signalling that appear to be deliberate rather than involuntary. Think of the bitch standing by the door, scratching at it and whining and looking back expectantly at her master as she tries to get him to let her out. Well, of course, learning theory explains this as a bit of conditioned behavior which can be accounted for without resort to talk of minds, or even the notion of trying. What about two dogs fighting? They snarl and posture threateningly, they maneuver for advantageous position, feint, lunge, and bite when they spot an opening, adjusting action from moment to moment to counter the opponent's moves and make mastering moves of their own. In cases like this, the moment-to-moment adjustments in the signals and stratagems are neither programmed as such by genes nor ordered as such by conditioning; they require perception, anticipation, perhaps even estimation of odds in the Maynard Smith manner. Only the behavioristic taboo holds back the natural inclination to view these actions as attempts to get the better of the opponent; that is to say, as attempts at manipulation in the literal sense of the word.

Manipulation in this sense implies intention; perceiving, anticipating, estimating are also intentional capacities. If, as Brentano and such modern philosophers as Boden (1972) and Midgley (1968) hold, intentionality is the mark of the mental, then there appear to be grounds for crediting a measure of mentality to animals.

Intentional System Theory

Recent developments in cognitive science have cast intentionality in another light, which might reflect doubt on whether it can stand as a premise for inference about animal minds. The artificial intelligence people have now gotten to the point of being able to program a computer so complexly that the only way for someone to deal with it at the surface level is to regard it as having beliefs, expectations, desires, reason, numerous of the intentional attributes. According to Daniel Dennett (1978), the only hope that someone has of beating one of the better chess-playing computers is to treat it as another chess player who figures out " . . . as best he can what the best or most rational move would be given the rules of the game" (Dennett, 1978, p. 5). Dennett has developed what he calls intentional system theory to deal with cases like this, in which a system can be viewed as rationally deciding on a course of action in pursuit of some goal, on the basis of information available to it and its understanding of the kind of situation with which it has to deal. Applied to machines, such as computers, he says that we should not get hung up on the question of whether the computer *really* has beliefs and desires and what not; the decision to adopt "the intentional stance" toward it is a pragmatic one—thinking of the machine as though it had beliefs and desires enables us to predict what it will do in situations where no other means of prediction is available or manageable.

However, some of the people in cognitive science appear to take a less hedged position, which John Searle (1980) refers to as "strong AI." This claims that an appropriately programmed computer *is* a mind in the full sense, and so literally understands, wants, visualizes, and judges things—simulation has passed over into duplication. If it could be accepted, therefore, strong AI would make an end of the difficulties Descartes created when he said that animals were merely machines, and also deal with the disjunctive dichotomy of Cartesian dualism. However, Searle (1980, 1982) has made a vigorous case against the claim, in which he argues that minds cannot be conceived of as computer programs, and hence that no machine so far invented has either mentality or the intentional states that go with it. If this is so, and if, at the same time, intentional idioms, with their full logical implications, can be used in talk about some modern computers, as Dennett claims, then it follows that using intentional idioms appropriately does not necessarily imply mentality. So the fact that we often ascribe intentionality to animals need not entail belief in an underlying mental life.

Ascribing intentionality to animals may, in certain cases, be the only means available to us to describe what we perceive them as doing. Just as such ascription can have heuristic value in the case of a machine, so too in the case of an animal it can make a certain kind of sense out of otherwise inexplicable behavior, and lead to our making testable predictions and asking answerable questions which we should otherwise be unlikely to think of. At least Dennett has argued that this might be so. At a recent Dahlem Conference on animal and human minds (Griffin, 1982), Dennett introduced Grice's (1957) notion of grades or orders of intentionality, and tried applying it to some animal cases (Seyfarth, 1982). To illustrate this notion, imagine a conversation in which Jack is trying to get Jill to believe that he wants her to believe that he loves her; or suppose that Jack is lying to Jill, knowing that she knows he is lying, his intent being to get her to think that he does not know that she knows. These examples comprise several orders of intentionality (more than one order of intentionality obtains if the intentional object of an intentional state is another intentional state). We are unlikely to find animals showing this degree of sophistication. At the other extreme is zero-order intentionality—no intentionality at all—such as when an involuntary scream of fear starts everyone running in panic. This seems more like the animal level. Well, suppose we take a case of an animal's alarm calling and interpret it this way. Then we should predict that an animal responding to presence of a predator by giving an alarm call will do so irrespective of whether it is aware of others of its kind in the vicinity. This can be tested by contriving a situation in which the animal encounters a predator when alone. If the animal then tries to slink silently away, instead of sounding the alarm, we should have some basis for promoting it to the first order of intentionality, which credits it, when giving the alarm call, with wanting to make its companions run for cover. We can suppose that they do so automatically on hearing the signal (just as one might blush on being shown an obscene picture). To test whether the flight response is blindly tuned in this way, the circumstances in which alarm calling occurs could be varied to give opportunity for discrimination to be shown. If, for example, the alarms of juveniles are ignored, and juveniles prove to be poor judges of what is a predator and what is not, we should have grounds for further intentional attribution. When Dennett and Robert Seyfarth applied this approach to the alarm calling of vervet monkeys, which have three different alarm calls, each for a different kind of predator, and three correspondingly different patterns of response (Seyfarth et al., 1980), they ended up with the monkeys on about the third rung of the intentionality ladder (Seyfarth et al., 1980).

Woodruff and Premak (1979) reported a case which they interpret as evidence that a chimpanzee can have a "theory of mind." The chimp in question sees food put in one of two boxes, and learns that if she points to that box her trainer will fetch the food and share it with her. Then another trainer comes on the scene, who keeps all the food for himself when the chimp points to the box containing it. After a few experiences of this, the chimp counters either by not pointing any more or by pointing at the empty box, when she has to deal with this selfish trainer. Woodruff and Premak concluded that, in the latter case, the ape was trying to deceive the selfish trainer, which implies that she thought that he would believe what she indicated to be her belief about the boxes. But could the ape not have been mindlessly repeating an act that had been reinforced, and so displaying no more than first-order intentionality? If its understanding had been any deeper, would it not have wondered why the selfish trainer had not gone to the right box on finding that the indicated one was empty? Or can it be supposed that the chimp assumed the man to be too stupid to think of that? Dennett and Sue Savage-Rambough suggested a way of probing the chimp's understanding further. If the chimp really were being rational it should give up its ruse as soon as it had reason to believe that the selfish trainer already knew where the food was. Putting the food in transparent boxes would create this condition, and so test the prediction (Seyfarth, et al., 1980). Thus intentional system theory can provide program for research.

Reciprocal altruism can also be looked at from this point of view. Indeed even before the Dahlem discussion of intentionalistic analysis of animal cases, John Crook (1980) had argued, in different terms, that reciprocal altruism requires at least third-order intentionality: A, the receiver of help, wants B, the helper, to believe that A will pay back the favor should the need arise. Likewise, ability to detect cheating, if it occurs in a literal as opposed to a Pickwickian sense, will involve high orders of intentionality.

It is understandable that the examples that most readily lend themselves to intentional analysis come from primates, the animals apparently closest to us in intellectual endowment. Yet I should not be surprised if more widespread application of the approach turns out to be feasible and productive, when more people catch on to the idea and refine the ways of putting it into practice. In my own studies of gulls, I now realize that I gradually and more or less unknowingly moved from a mechanistic to an intentionalistic conception of the birds and their behavior, especially after stumbling on the fact that they recognize one another individually, and manifest this recognition in a variety of different ways at different times and in different contexts (Beer, 1973, 1975). Observations and experiments on vocal interactions between Laughing Gull parents and chicks at different stages of family life revealed a phase during which obedience to vocal summons ceases to be the rule, and the parent and chick appear to engage in a contest of wills over who will approach whom (Beer, 1979). For the young gull, this may well furnish experience necessary for acquisition of competence in many of the kinds of communication skill that have to be mastered by maturity. Early and late the notion of manipulation appears literally to apply. In playback experiments, responsiveness to recorded calls, in addition to demonstrating individual recognition by ear, showed waning when birds apparently perceived a lack of timing relationship between the calls they were giving and those they were hearing; and there was also variation in response to the playback depending upon whether the individuals from which the calls were recorded were present or absent during a test. The gulls thus showed themselves to be aware of and attentive to far more of the detail in social situations, than the older mechanistic, probabilistic, and instinct theories of animal social communication (e.g., Tinbergen, 1951, 1953) predisposed one to look for.

Conclusion

If it be conceded that the description of animal behavior involves and may even require the use of intentional terms, what conclusions of an ethical sort follow? We have seen that such use does not necessarily imply belief in animal mentality or sentience, since it can be applied to machines without our having to believe that the machines are possessed of minds. Intentionality may be a necessary condition of mind, but not a sufficient condition. But even if it were true that intentionality is the mark of the mental, for animals as well as for people, the ethical consequences would remain questionable. Mentality may be a necessary condition to qualify a creature for humane consideration, but it is not a sufficient condition, as man's treatment of man makes plain. It is too late in my paper to try to tackle this matter, and I lack the competence to do so anyway. However, I am persuaded that our attitudes toward animals, and how we treat them, reflect and are reflected by how we talk about them. Conception and language are intimately intertwined and reciprocally influence one another. The conception of the animal as a machine goes with causal talk, which tends to have no place for ethical considerations. The conception of the animal as an intentional agent goes with intentional talk, which can encompass notions of well-being and suffering, which open the way to questions about rights and interests.

So in answer to Huxley's question about whether we have reason to think that animals are anything but insentient automata, I echo Forster's old lady: I shall know what I think when I see what I say; but I cannot pretend to have no idea of what that is likely to be.

References

Auden, W. H. (1968). *The Dyer's Hand*. Vintage Books, New York.

Beer, C. G. (1973). A view of birds. *Minnesota Symp. Child Psychol.* **7**, 47–86.

Beer, C. G. (1975). Multiple functions and gull displays. In: *Function and Evolution in Behaviour—Essays in Honour of Professor Niko Tinbergen*. (Baerends, G., Beer, C., and Manning, A., eds.) pp. 16–54. Clarendon Press, Oxford.

Beer, C. G. (1979). Vocal communication between Laughing Gull parents and chicks. *Behaviour* **70**, 118–146.

Black, M. (1962). *Models and Metaphors*. Cornell University Press, Ithaca, New York.

Boden, M. (1972). *Purposive Explanation in Psychology*. Harvard University Press, Cambridge, Massachusetts.

Boden, M. (1979). The computational metaphor in psychology. In: *Philosophical Problems in Psychology* (Bolton, N., ed.), pp. 111–132. Methuen, London.

Chisholm, R. M. (1967). Intentionality. In: *The Encyclopedia of Philosophy* (Edwards, P., ed.), Vol. 4, pp. 201–204. Macmillan and Free Press, New York.

Crook, J. H. (1980). *The Evolution of Human Consciousness*. Clarendon Press, Oxford.

Dawkins, R. (1976). *The Selfish Gene*. Oxford University Press, Oxford.

Dawkins, R., and Krebs, J. R. (1978). Animal signals: Information or manipulation? In: *Behavioural Ecology—An Evolutionary Approach* (Krebs, J. R. and Davies, N. B., eds.), pp. 282–309. Blackwell, Oxford.

Dennett, D. C. (1969). *Content and Consciousness*. Routledge & Kegan Paul, London.

Dennett, D. C. (1978). *Brainstorms*. Bradford Books, Montgomery, Vermont.

Grice, H. P. (1957). Meaning. *Philosoph. Rev.* **66**, 377–388.

Grice, H. P. (1969). Utterer's meaning and intentions. *Philos. Rev.* **78**, 147–177.

Griffin, D. R. (1976). *The Question of Animal Awareness*. Rockefeller University Press, New York.

Griffin, D. R. (ed.) (1982). *Animal Mind—Human Mind*. Springer-Verlag, Berlin.

Humphrey, N. K. (1976). The social function of intellect. In: *Growing Points in*

Ethology (Bateson, P. P. G., and Hinde, R. A. eds.), pp. 303–317. Cambridge University Press, Cambridge.

Huxley, T. H. (1893). On the hypothesis that animals are automata, and its history. In: *Method and Results: Essays*. John Murray, London.

Maynard Smith, J. (1974). The theory of games and the evolution of animal conflicts. *J. Theoret. Biol.* **47**, 209–221.

Maynard Smith, J. (1976). Evolution and the theory of games. *Am. Sci.* **64**, 41–45.

Midgley, M. (1978). *Beast and Man*. Cornell University Press, Ithaca.

Midgley, M. (1979). Gene-juggling. *Philosophy* **54**, 439–458.

Packer, C. R. (1977). Reciprocal altruism in *Papio anubis. Nature (London)* **265**, 441–443.

Searle, J. R. (1980). Minds, brains and programs. *Behav. Brain Sci.* **3**, 417–424.

Searle, J. R. (1982). The myth of the computer (Review of D. R. Hofstadter and D. C. Dennett, *The Mind's I*. Basic Books, New York.). *N. Y. Rev. Books* (April, 29, 1982), **29**, No. 7, 3–6.

Seyfarth, R. M. (1982). Communication as evidence of thinking. In: *Animal Mind—Human Mind* (Griffin, D. R., ed.), pp. 391–404. Springer-Verlag, Berlin.

Seyfarth, R. M., Cheney, D. L., and Marler, P. (1980). Vervet monkey alarm calls: Semantic communication in a free-ranging primate. *Animal Behav.* **28**, 1070–1094.

Taylor, C. (1964). *The Explanation of Behaviour*. Routledge & Kegan Paul, London.

Taylor, G. R. (1963). *The Science of Life*. Thames & Hudson, London.

Tinbergen, N. (1951). *The Study of Instinct*. Clarendon Press, Oxford.

Tinbergen, N. (1953). *The Social Behaviour of Animals*. Methuen, London.

Trivers, R. L. (1971). The evolution of reciprocal altruism. *Quart. Rev. Biol.* **46**, 35–57.

Urmson, J. O. (1956). *Philosophical Analysis—Its Development between the Wars*. Clarendon Press, Oxford.

Urmson, J. O., and Cohen, J. (1968). Criteria of intensionality. *Proc. Aristotelian Soc.* (Suppl.) **42**, 107–142.

Wittgenstein, L. (1922). *Tractatus Logico-philosophicus*. Kegan Paul, London.

Wittgenstein, L. (1958). *Philosophical Investigations*, 2nd ed. Blackwell, Oxford.

Woodruff, G. and Premack, D. (1979). Intentional communication in the Chimpanzee—the development of deception. *Cognition* **7**, 333–362.

Chapter 10

Neurobiological Origins of Human Values

DONALD W. PFAFF*

In reviewing the special ethical problems of treating patients with disorders of the nervous system or behavior, we encounter conflicts between values. Ruth Macklin (1982, 1983a,b) assures us that a philosopher expects many patient-care situations to involve built-in tensions among values. Frequently, for example, doctors will have to choose among standards which tend to increase the autonomy of the patient *versus* those which maximize the doctors' expected medical benefit to the patient. Most importantly, Macklin (1983a,b) says, there are no philosophically final decision rules for choosing among values: There exists no purely philosophical, rational approach to logically closed arguments in these situations of value conflict.

Is there a role for scientific approaches to problems of the justification of values? While many would talk about is–ought distinctions and claim that scientific explanations must be restricted to matters of fact as opposed to values, others, notably Sperry (1977), sense that the neural and behavioral sciences may contribute fundamentally to the understanding of values. That is, ethical behavior is a factual matter susceptible to scientific explanation. Sperry also says (1972) that such understanding is important because behavioral manifestations of human values, for example, in overpopulation and unlimited aggression, are the underlying causes of the difficulties most

*Laboratory of Neurobiology and Behavior, The Rockefeller University, New York, New York 10021.

threatening to our species. Eventually, a scientist like Sperry thinks, a scientific understanding of our value system will allow us as a civilization to check it regularly and see how it fits with the requirements for survival under given conditions of the physical world.

In the abstract, the most exciting possibility is that neuroscience offers the opportunity to construct logically closed arguments which rationalize our choices among values. That is, we may be able to understand some of the mechanisms involved in the ethical aspects of an ethical act, and based especially on the limitations of our nervous system, understand why we are and should be the ethical creatures we are. In the absence of an *a priori* philosophically closed argument for decisions among values, we would have self-consistent and internally complete arguments based on combinations of philosophy and neuro-biological understanding.

Justifications of human morality have not always remained the same. During the Renaissance, ethics based purely on Christian theology found new justification in a lay morality. In the next few decades, scientific understanding may reach such a point as to contribute to a biologically based morality.

One scientific approach to the understanding of human values has come from evolutionary thought (Wilson, 1975, 1979). Jerram Brown (1983) gave examples of cooperation between members of a species across a wide variety of animals. One type of explanation for altruistic behavior is natural selection at the level of the group rather than the individual (Wilson, 1975). But, in his chapter, Arthur Caplan (1983) warned us of limits on that sort of sociobiological thought which would extrapolate from animal behavior to human behavior and say that cultural values are epigenetic expressions of characteristics which evolved as a result of group selection (Lumsden and Wilson, 1981). One of those limitations was the possiability of convergent evolution: The reason a human does something may be quite different from the reason an analogous behavior appeared in another species. A second limitation is that some of our behavioral characteristics, including ethical choices, may be neutral with respect to natural selection.

The other way of using biological knowledge to shed light on our use of values is through neurobiological mechanisms: When will it be possible to understand, neurophysiologically, how the ethical aspect of an ethical behavior is generated? For such an ambitious question, it certainly is wisest to follow the strategy of looking for ethical universals. This is because, on first thought, we are most likely to encounter anthropologically stable, solid, discoverable mechanisms for those ethical features which always occur in the species. So, our goal is a feasibility argument for a neurobiological understanding of an ethical universal. This would be a generalizable, justifiable rule for guiding action toward other people (Little and Twiss, 1978).

An Ethical Universal

One rule clearly qualifies as a candidate for being an ethical universal: the requirement for reciprocal cooperation. A first step is to document the existence of this rule in a wide variety of manifestations in human history. Consider the ethical aspects of several religions as examples of collected wisdom in various human cultures. Following are quotations from a wide variety of religions (Dās, 1947), which illustrate the wide historical scope of what, in Christianity, would come to be called the Golden Rule.

From Oriental origins, quotations include:

> A disciple asked the Chinese Master 'Is there one word which may serve as a rule of practice for all one's life?' Confucius answered: 'Is *reciprocity* not such a word? Do not to others what you do not want done to yourself—this is what the word means.' (Confucius, Analects, 15.23)

> Pity the misfortunes of others; rejoice in the well-being of others; help those who are in want; save men in danger; rejoice at the success of others; and sympathize with their reverses, *even as though you were in their place.* (Taoist writings, Tai-Shang-Kan-Ying-Pien)

From orthodox Hindu philosophy:

> Do not to others what ye do not wish done to yourself; and wish for others too what ye desire and long for, for yourself—This is the whole of Dharma, heed it well. (Vedic Scriptures, Mahā-bhārata)

From a Persian religion, centered around Zoroaster, the prophet:

> That which is good for all and any one, for whomsoever—that is good for me . . . What I hold good for self, I should for all. Only Law Universal is true Law. (Zoroastrian writings, Gāthā, 43.1)

From the 6th century B.C., in India:

> Harmlessness is the only religion. (Jain maxim, Ahimsa Paramo Dharmah)

From the Old Testament:

> Thou shalt love God above all things, and thy neighbor as thyself. (Leviticus)

From the New Testament, the Christian Golden Rule:

'Teacher, which is the great commandment in the law?' And he said to him, 'You shall love the Lord your God with all your heart, and with all your soul, and with all your mind.' This is the great and first commandment. And a second is like it, you shall love your neighbor as yourself. On these two commandments depend all the law and the prophets. Matthew 22:36–40

And, two examples from Islam:

Noblest religion this—that thou shouldst like for others what thou likest for thyself; and what thou feelest painful for thyself, hold that as painful for all others too. (Hadis, Sayings of Muhammad)

Noblest religion this—that others may feel safe from thee; the loftiest Islām—that all may feel safe from thy tongue and hands. (Qurān)

The striking omission is Buddhism, whose teaching is consistent with the Golden Rule but precludes a direct statement of it. The postulate of a universal spirit, Anatman, says that each of our human existences is not separate from this universal spirit (Dās, 1947), and that when any of us reaches a deep meditative state we realize that. Since each of us as an individual is part of the same thing, we would hardly take actions to harm another part of that universal existence. On the other hand, a direct statement of that implied reciprocity is impossible because in the realization of a univeral existence the separate individual words for "I" and "you" do not make sense. Indirect statements amounting to the Golden Rule also appear among the Navajo, American Indians (Kluckhohn, 1949).

More recently, and outside an explicit religious context, is the categorical imperative of Immanuel Kant.

There is, therefore, only one categorical imperative. It is: Act only according to that maxim by which you can at the same time will that it should become a universal law. (Kant, Foundations of the Metaphysics of Morals)

Reciprocity is directly implied here, since in harming another and generalizing from his action, a person would be allowing the other to harm him.

Game Theory

The human applications of the Golden Rule, or the categorical imperative, cross continents and centuries, but another example of its universality comes from a computer-programmed application of game theory. Axelrod and Hamilton (1981) in an article in *Science* wanted to

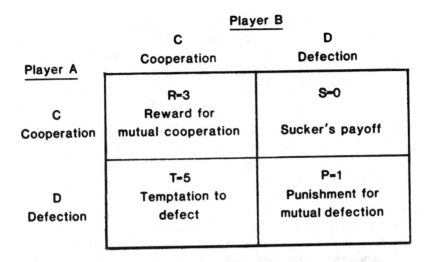

Figure 10.1. A two-player version of the Prisoner's Dilemma game. Numbers are the payoff to player A in terms of his fitness for survival. This game is defined by $T > R > P > S$ and $R > (S + T)/2$. On repeated encounters, both players do better cooperating than acting selfishly ("defecting") (from Axelrod and Hamilton, 1981).

account for the existence of cooperation and altruism in a mathematically described way. Where cooperating individuals are closely related genetically, altruism can benefit reproduction of the set (Wilson, 1975), so that genes for cooperative behavior will be passed on even if a cooperating individual is sacrificed. But Axelrod and Hamilton wanted to consider, in a precise way, conditions under which reciprocal cooperation could evolve without regard to how closely related the cooperating individuals are. They used the game of Prisoner's Dilemma to construct a formal theory of cooperation. Figure 10.1 shows their two-player version of the game, in which the two individuals can cooperate or defect (act selfishly). The numerical payoff to a player is in terms of the effect of each encounter on his survival. In Figure 1, the payoff from the point of view of player A is shown. The game is set up so that from the selfish point of view of player A, whichever player B does, player A has a higher reward for acting selfishly than for cooperating. But if both act selfishly, they do worse than if both had cooperated. The key point is that if two individuals in this game have just one encounter, player A clearly should defect. So it is crucial that in a wide variety of human interactions, the same individuals will meet many times. Therefore, the program for the game includes the assumption that there will be a high probability that the same two individuals will meet again. Axelrod and Hamilton arranged a computer tournament for strategies to be followed by the two individuals in this Prisoner's Dilemma game, each strategy to be applied for a game length of 200 moves. The highest average score was reached by the following strategy: that each player should cooperate on the first move and then do whatever the other player did on the preceding move. This is a strategy of cooperation based on reciprocity.

The circumstances of Axelrod's and Hamilton's (1981) computer tournament showed that the strategy of reciprocal cooperation is *robust*. It won against a range of strategies that included some very sophisticated ones. By asking about *stability*, Axelrod and Hamilton required that if the strategy of reciprocal cooperation was established, it would survive even if a small part of the population started using a different one. At some length, Axelrod and Hamilton prove that for individuals who have a large probability of meeting again the best of the possible invading strategies could not achieve a higher score than it does. Finally, they ask how the strategy of reciprocal cooperation could get started (*initial viability*). Suppose you have a population of individuals who always act selfishly. Axelrod and Hamilton show that there are at least two mechanisms by which reciprocal cooperation could get started and then thrive. The first mechanism is if a small number of individuals in the population who are genetically related to each other begin to cooperate. They can do this because it essentially amounts to a recalculation of the payoff matrix for the game: An individual player has a part interest in the other player's gain because he is related to the other player. The other mechanism is clustering. If a small number of inidividuals in the population have a high proportion of their interactions with each other, mathematically they can reciprocally cooperate with profit. Clustering could be a result of kinship but does not have to be.

Thus, without respect to kinship or religion, Axelrod and Hamilton (1981) show abstract game conditions where the programmed strategy of cooperation based on reciprocity can be proven to be intially viable, competitively robust, and stable against invasions by other strategies. Not only does the mathematical formulation of this part of game theory show a further breadth of reciprocal cooperation, but also the susceptibility of this ethical universal to a formulation using precise programming suggests a similar susceptibility to eventual biologically mechanistic analysis.

Neurobiological Approaches

How could one account for an ethical act, neurophysiologially? Of course, at this stage of development of the neurosciences one can hope

Table 10.1.   Steps in Obeying an Ethical Universal (Reciprocity)

Consider an action toward another individual

1. Represent action in CNS.
2. Remember its consequences.
3. Associate them with yourself.
4. If positive or neutral action OK. If negative, action not OK.

only for a feasibility argument. But the exercise is worthwhile if only to point out that some popular approaches to this question may be wrong. Usually in discussion of those aspects of human behavior that could involve ethics, words like "rational," "responsible," "idealism," "reason," "think," and "symbolize" are used. For such sophisticated formulations it would be natural, in the neurologizing by MacLean, for example, to assign responsibility for the ethical aspects of ethical behavior only to the most recently evolved neural mechanisms (MacLean, 1949, 1967). In MacLean's triune brain, one has the lowest type of neural mechanism, governing reflex and vegetative aspects of behavior, the reptilian brain, which esentially adds limbic telencephalon and, most recently, the neomammalian brain, which essentially is dorsolateral neocortex. Applying this division to higher mental activity, MacLean emphasizes that our emotional behavior remains a function of crude, primitive neural systems and implies that behavior conforming to a moral principle would require domination of our reptilian brain by sophisticated, recently evolved neocortical tissue.

A Parsimonious Model

It is more useful for the sake of envisioning possible neurophysiological explanations to be parsimonious, in the minimal description of a behavior which obeys an ethical universal, and to avoid the unnecessary use of mentalistic terminology. When a person is about to act toward another peson he must do the following things (Table 10.1) so that the resulting behavior will fit the requirements for reciprocal cooperation: Represent the action, remember its consequences, and evaluate them as though they were applied to himself. The requirement for *representing an action in the central nervous system before it occurs* need not involve neural mechanisms which are special for ethics. Following the work of von Holst, Held and Freedman (1965) accounted for the perceptual fact that the three-dimensional world appears stable even when we move. The experiments and concept they used included the postulate of a "reafferenz signal"—a representation of the motor signal separate from that which would excite skeletal muscles, which would inform afferent systems of the movement which would occur. Similarly, Teuber (1960) postulated a "corollary discharge"—a signal representing, for example, eye movements, such that the visual world would not appear to jump when an organism's eyes move about. These precedents for representation of a behavior may be related to well-documented electrophysiological examples of cerebral cortex potentials preceding a voluntary movement. Vaughan *et al.* (1968) recorded a negative potential over the motor cortex of human subjects as long as 2 sec before voluntary wrist movement. This motor potential was not dependent on kinesthetic afferents (Vaughan and Gross, 1970), and has been analyzed further in monkeys by comparing surface recordings with those from electrodes implanted in the cortex (Arezzo and Vaughan, 1975). Similarly, Kornhuber and Deecke (1965) found surface-negative potentials before limb movements, over the contralateral precentral cortex. This "readiness potential" was enhanced by attention (Deecke and Kornhuber, 1977). Cortical electrical changes preceding movement have been analyzed more discretely by single unit recording from antidromically identified pyramidal cells (Evarts, 1972, 1974). Under certain conditions of training, cortical neurons in the monkey cortex whose discharge is associated with a certain action may show elevated firing rates long before that action occurs (Evarts and Tanji 1976; Tanji *et al.*, 1980).

Consider the remaining steps in a behavior which obeys the requirement for reciprocal cooperation. *Remembering the consequences of the act* is nothing more than simple associative memory, mechanisms for which are being studied in a wide variety of species. The formulation *"associate consequences with yourself"* could sound vague, but equally well could represent the simplest step. That is, failing to distinguish the conspecific from yourself involves simply forgetting whether, on previous occurrences of that act, the other person or yourself would be harmed. Finally, the evaluative step says that *if the consequences are positive the act is ethical; if negative, not.* This need be nothing more than the mechanisms of reinforcement that are studied by physiologists and neurochemists in a wide variety of species. For example, if sudden rises in blood pressure were part of the negative reinforcement associated with the memory of pain, when that increase in blood pressure and other signals of negative reinforcement occurred, the potential behavioral act would not occur.

The complexities in the neurophysiological explanations for this series of steps are not a function of the ethical aspects of the behavior but, for example, in the explanation of the sensory and motor contingencies of the act when they are complicated. Similarly, understanding the mechanisms of positive and negative reward, especially with physiological systems such as blood presure (itself intricately regulated), presents a major challenge. Thus, except where motor acts which require neocortex for their very execution are involved, ethical behavior may consist of a series of relatively primitive steps, in which, especially in their association with positive or negative reward, neurologically primitive tissue in the limbic system and brainstem, play the crucial roles.

This argument holds out the possibility that ethical behavior may not be any more difficult to understand physiologically than many of the other functions neuroscientists try to understand. In other contexts, considerable progress has been made toward elucidating neurophysiological mechanisms for a mammalian reproductive behavior (Pfaff, 1980), which include a mechanism of motivation (Pfaff, 1982). The series of physiologically realizable steps envisioned for reciprocal cooperation does assume that the organism is familiar with the act considered. This implies experience with and memory for a large variety of motor acts. In fact, limitations on an organism's capacity for ethical peformance may derive from limited perceptual and memory abilities (Beer, 1983). The sequence also assumes the organism is familiar with the consequences of the act. This implies that individuals may show decreased responsibility for acts with distant consequences which are and have been difficult to appreciate. The argument also suggests that social learning toward representing an act before it occurs (that is, avoiding impulsive behavior) would help in allowing a person to enter the sequence which yields reciprocally cooperative behavior.

Eventual Usefulness

It would be a surprise if a type of ethical act usually held to be among the highest expressions of human culture could actually be found to depend upon neural mechanisms which are rather primitive. Indeed, some philosophers (Westermarck, 1912) have emphasized the *emotional* nature of moral judgments. To a neuroscientist this is interesting for its own sake, and an exercise done as above helps to show where the real technical difficulties in explaining such complex behaviors lie. But neurophysiological understanding may also be important for another reason. By analogy, in automata theory one can speak of the behavior of the machine as moving through allowed sequences of locations as programmed in an n-dimensional space. Since many of the automata considered are not only designed by humans but simple enough to explain completely, the engineer can both elucidate the sequences of states that represent the normal operation of the machine and point out what we might call the "pathologies." For example, automata can have

states from which, once they've entered, they can never emerge: these
are "trap states." In other cases, locations in the space could be
disallowed because the electronics of the machine simply can not reach
them or because, if those states were entered, the machine would be
damaged. If, for this analogy, human beings can be thought of as
"ethical automata," when will we understand our neural machinery well
enough to know which states should be disallowed? Particularly as we
face tensions between competing values, as often is the case in patient
care, we want to consider our value system. Adequate understanding of
the neurobiological mechanisms underlying ethical behavior will be
useful at least for pointing out the built-in dangers of trying to change
it.

References

Arezzo, J., and Vaughan, H. G., Jr. (1975). Cortical potentials associated with
  voluntary movements in the monkey. *Brain Res.* **88**, 99–104.
Axelrod, R., and Hamilton, W. D. (1981). The evolution of cooperation. *Science*
  211, 1390–1396.
Beer, C. (1983). Motives and metaphors in consideration of animal nature. In:
  Ethical Questions in Brain and Behavior: Problems and Opportunities.
  Springer-Verlag, New York, this volume, pp. 125–140.
Brown, J. (1983). Some paradoxical goals of cells and organisms: The role of
  MHC. In: *Ethical Questions in Brain and Behavior: Problems and Oppor-
  tunities.* Springer-Verlag, New York, this volume, pp. 111–124.
Caplan, A. (1983). Out with the "old" and in with the "new"—The evolution
  and refinement of sociobiological theory. In: *Ethical Questions in Brain and
  Behavior: Problems and Opportunities.* Springer-Verlag, New York, this
  volume, pp. 91–109.
Dās, B. (1947). *The Essential Unity of all Religions.* 3rd ed. Ananda Publishing
  House, Benares.
Deecke, L., and Kornhuber, H. H. (1977). Cerebral potentials and the initiation
  of voluntary movement. In: *Attention, voluntary contraction and event-related
  cerebral potentials.* Progress in Clinical Neurophysiology (Desmedt, J. E.,
  ed.), Vol. 1. Basel, Karger.
Evarts, E. V. (1972). Contrasts between activity of precentral and postcentral
  neurons of cerebral cortex during movement in the monkey. *Brain Res.* **40**,
  25–31.
Evarts, E. V. (1974). Precentral and postcentral cortical activity in association
  with visually triggered movement. *J. Neurophysiol.* **37**, 373–381.
Evarts, E. V., and Tanji, J. (1976). Reflex and intended responses in motor
  cortex pyramidal tract neurons of monkey. *J. Neurophysiol.* **36**, 1069–
  1080.
Held, R., and Freedman, S. J. (1963). Plasticity in human sensorimotor control.
  Science **142**, 455–462.

Kluckhohn, C. (1949). The philosophy of Navajo indians. In: *Ideological Differences and World Order* (Northrop, F. S. C., ed.). Yale University Press, New Haven, Connecticut.

Kornhuber, H. H., and Deecke, L. (1965). Hirnpotentialänderungen bei Willkürbewegungen und passiven Bewegungen des Menschen: Bereitschaftspontential und reafferente Potentiale. *Pflügers Arch.* **284**, 1–17.

Little, D., and Twiss, S. B., Jr. (1978). *Comparative Religious Ethics.* Harper & Row, New York.

Lumsden, C., and Wilson, E. O. (1981). *Genes, Mind and Culture.* Harvard University Press, Cambridge, Massachusetts.

Macklin, R. (1982). *Man, Mind and Morality. The Ethics of Behavior Control.* Prentice-Hall, Englewood Cliffs, New Jersey.

Macklin, R. (1983a). Problems of informed consent with the cognitively impaired. In: *Ethical Questions in Brain and Behavior: Problems and Opportunities.* Springer-Verlag, New York, this volume, pp. 23–40.

Macklin, R. (1983b). Treatment refusals: autonomy, paternalism, and the "best interest" of the patient. In: *Ethical Questions in Brain and Behavior: Problems and Opportunities.* Springer-Verlag, New York, this volume, pp. 41–56.

MacLean, P. D. (1949). Psychosomatic disease and the visceral brain. *Psychosom. Med.* **11**, 338–353.

MacLean, P. D. (1967). The brain in relation to empathy and medical education. *J. Nerv. Mental Dis.* **144**, 374–383.

Pfaff, D. W. (1980). *Estrogens and Brain Function: Neural Analysis of a Hormone-Controlled Mammalian Reproductive Behavior.* Springer-Verlag, New York.

Pfaff, D. W. (ed.) (1982). *The Physiological Mechanisms of Motivation.* Springer-Verlag, New York.

Sperry, R. W. (1972). Science and the problem of values. *Perspect. Biol. Med.* **16**, 115–130.

Sperry, R. W. (1977). Bridging science and values. A unifying view of mind and brain. *Am. Psychologist* **32**, 237–245.

Tanji, J., Taniguchi, K., and Saga, T. (1980). Suplementary motor area: Neuronal response to motor instructions. *J. Neurophysiol.* **43**, 60–68.

Teuber, H. -L. (1960). Perception. In: *Handbook of physiology. Section 1: Neurophysiology* (Field, J., Magoun, H. W., Hall, and V. E., eds.), Vol. III. American Physiolial Society, Washington, D. C.

Vaughan, H. G., Jr., Costa, L. D., and Ritter, W. (1968). Topography of the human motor potential. *Electroencephalogr. Clin. Neurophysiol.* **25**, 1–10.

Vaughan, H. G., Jr., and Gross, E. G. (1970). Cortical motor potential in monkeys before and after upper limb deafferentation. *Exp. Neurol.* **26**, 253–262.

Westermarck, E. A. (1912). *The Origin and Development of Moral Ideas.* MacMillan, London.

Wilson, E. O. (1975). *Sociobiology.* Harvard University Press, Cambridge, Massachusetts.

Wilson, E. O. (1979). *On Human Nature.* Harvard University Press, Cambridge, Massachusetts.

Index